CARBON CONFLICTS AND FOREST LANDSCAPES IN AFRICA

Amidst the pressing challenges of global climate change, the last decade has seen a wave of forest carbon projects across the world, designed to conserve and enhance forest carbon stocks in order to reduce carbon emissions from deforestation and offset emissions elsewhere. Exploring a set of new empirical case studies, *Carbon Conflicts and Forest Landscapes in Africa* examines how these projects are unfolding, their effects and who is gaining and losing. Situating forest carbon approaches as part of more general moves to address environmental problems by attaching market values to nature and ecosystems, it examines how new projects interact with forest landscapes and their longer histories of intervention. The book asks: what difference does carbon make? What political and ecological dynamics are unleashed by these new commodified, marketized approaches, and how are local forest users experiencing and responding to them?

The book's case studies cover a wide range of African ecologies, project types and national political–economic contexts. By examining these cases in a comparative framework and within an understanding of the national, regional and global institutional arrangements shaping forest carbon commoditisation, the book provides a rich and compelling account of how and why carbon conflicts are emerging, and how they might be avoided in future.

This book will be of interest to students of development studies, environmental sciences, geography, economics, development studies and anthropology, as well as practitioners and policy makers.

Melissa Leach is Director of the Institute of Development Studies, University of Sussex, UK.

Ian Scoones is a Professorial Fellow at the Institute of Development Studies, University of Sussex, UK, and co-directs the ESRC STEPS (Social, Technological and Environmental Pathways to Sustainability) Centre, UK.

Carbon forestry is privatizing, commodifying and financializing the world's forests, recasting relations between state and market in forest landscapes. This book illuminates the fraught political economy of this transformative moment – through lived experience within place-based histories. As the first comparative political ecology of carbon forestry politics, this book is essential reading for scholars and practitioners wishing to transform carbon forestry for the better.

Jesse Ribot,
University of Illinois, USA

This book not only synthesizes what we know about carbon forestry and illustrates how it has unfolded in Africa, it also critically reflects on the material, social and cultural life of carbon and how the latter features amidst dynamic ecologies and the development needs and aspirations of states and people. This is a brilliant book: a must read for scholars and activists interested in the commodification of environmental services and their likely consequences.

Esteve Corbera,
Universitat Autònoma de Barcelona, Spain

This book will help readers to better understand why it is important to incorporate livelihood considerations and a landscape approach into the design and implementation of forest carbon projects.

Gretchen Walters,
International Union for the Conservation of Nature (IUCN), Switzerland

Pathways to Sustainability Series

This book series addresses core challenges around linking science and technology and environmental sustainability with poverty reduction and social justice. It is based on the work of the Social, Technological and Environmental Pathways to Sustainability (STEPS) Centre, a major investment of the UK Economic and Social Research Council (ESRC). The STEPS Centre brings together researchers at the Institute of Development Studies (IDS) and SPRU (Science and Technology Policy Research) at the University of Sussex with a set of partner institutions in Africa, Asia and Latin America.

Series Editors:
Ian Scoones and Andy Stirling
STEPS Centre at the University of Sussex

Editorial Advisory Board:
Steve Bass, Wiebe E. Bijker, Victor Galaz, Wenzel Geissler, Katherine Homewood, Sheila Jasanoff, Melissa Leach, Colin McInnes, Suman Sahai, Andrew Scott

Titles in this series include:

Dynamic Sustainabilities
Technology, environment, social justice
Melissa Leach, Ian Scoones and Andy Stirling

Avian Influenza
Science, policy and politics
Edited by Ian Scoones

Rice Biofortification
Lessons for global science and development
Sally Brooks

Epidemics
Science, governance and social justice
Edited by Sarah Dry and Melissa Leach

Regulating Technology
International Harmonization and Local Realities
Patrick van Zwanenberg, Adrian Ely, Adrian Smith

The Politics of Asbestos
Understandings of Risk, Disease and Protest
Linda Waldman

Contested Agronomy
Agricultural research in a changing world
James Sumberg and John Thompson

Transforming Health Markets in Asia and Africa
Improving quality and access for the poor
Edited by Gerald Bloom, Barun Kanjilal, Henry Lucas and David H. Peters

Pastoralism and Development in Africa
Dynamic change at the margins
Edited by Ian Scoones, Andy Catley and Jeremy Lind

The Politics of Green Transformations
Ian Scoones, Melissa Leach and Peter Newell

Carbon Conflicts and Forest Landscapes in Africa
Edited by Melissa Leach and Ian Scoones

CARBON CONFLICTS AND FOREST LANDSCAPES IN AFRICA

Edited by
Melissa Leach and Ian Scoones

Routledge
Taylor & Francis Group
LONDON AND NEW YORK

earthscan
from Routledge

First published 2015
by Routledge
2 Park Square, Milton Park, Abingdon, Oxon OX14 4RN

and by Routledge
711 Third Avenue, New York, NY 10017

Routledge is an imprint of the Taylor & Francis Group, an informa business

British Library Cataloguing in Publication Data
A catalogue record for this book is available from the British Library

Library of Congress Cataloging in Publication Data
A catalog record has been requested

ISBN: 978-1-138-82482-9 (hbk)
ISBN: 978-1-138-82483-6 (pbk)
ISBN: 978-1-315-74041-6 (ebk)

Typeset in Bembo
by Saxon Graphics Ltd, Derby

CONTENTS

FIGURES AND TABLES

List of figures (titles and sources – where sources are not given they are original, and developed by the author)

List of tables (titles and sources – where sources are not given they are original, and developed by the author)

ACRONYMS AND ABBREVIATIONS

A/R	Afforestation and Reforestation
A/R CDM	Afforestation/Reforestation Clean Development Mechanism
ABMS	Activity Baseline Monitoring Survey
ADC	Area Development Committee
AFOLU	Agriculture, Forestry and Other Land Use
AFP	Absentee Farmer Programme
ASAL	Arid and Semi-Arid Land
BCP	BioCarbon Partners
BioCF	BioCarbon Fund
CBO	Community Based Organization
CCAFS	CGIAR Research Programme on Climate Change, Agriculture and Food Security
CCB	Climate, Community and Biodiversity Standard
CCBA	Climate, Community and Biodiversity Alliance
CCP	Carbon Credit Project
CDM	Clean Development Mechanism
CERs	Certified Emission Reductions
CFA	Community Forest Association
CFR	Central Forest Reserve
CMP	Meeting of the Parties to the Kyoto Protocol
COMPACT	Community Management of Protected Areas Conservation Trust
COP	Conference of Parties
DDCC	District Development Coordinating Committee
DIP	Decentralization Implementation Plan
DNA	Designated National Authority

DOE	Designated Operational Entities
EB	CDM Executive Board
EDC	Environmental Development Consultants
EMC	Environmental Management Committee
EMCA	Environmental Management and Coordination Act
ENFORAC	Environmental Forum for Action
EPA-SL	Environmental Protection Agency, Sierra Leone
ERD	Environment and Rural Development
ES	Ecosystem service
ETS	Emissions Trading Schemes
FAO	United Nations Food and Agriculture Organization
FCPF	Forest Carbon Partnership Facility
FD	Forestry Department
FFU	Field Force Unit
FPIC	Free Prior Informed Consent
FSC	Forest Stewardship Certification
GEF	Global Environmental Facility
GHG	Greenhouse gas
HURINET	Human Rights Network
ICRAF	The World Agroforestry Centre
IFC	International Finance Corporation
JFM	Joint Forest Management
JI	Joint Implementation
KACP	Kenya Agricultural Carbon Project
KARI	Kenya Agricultural Research Institute
KCRP	Kariba Carbon REDD Project
KINAPA	Kilimanjaro National Park Authority
KWS	Kenya Wildlife Service
LULUCF	Land use, land-use change and forestry
LZRP	Lower Zambezi REDD+ Project
MAFFS	Ministry of Agriculture, Forestry and Food Security
MDG	Millennium Development Goals
MHCA	Maungu Hills Conservancy Association
MLCPE	Ministry of Lands, Country Planning and the Environment
MRV	Monitoring, Reporting and Verification
NFA	National Forestry Authority
NFC	New Forests Company
NGO	Non-governmental organization
NTFP	Non-timber forest product
OBF	Österreichische Bundesforste
OTC	Over-the-counter Market
PaViDIA	Participatory Village Development
PDD	Project Design Document
PES	Payments for Ecosystem Services

RAS	Regional Administrative Secretary
RC	Regional Commissioner (Kilimanjaro Region)
RDC	Rural District Council
REDD+	Reducing Emissions from Deforestation and Degradation plus the role of conservation, sustainable forest management and carbon enhancement
RoSL	Republic of Sierra Leone
SADC	Southern Africa Development Community
SCC-ViA	Swedish Cooperative Centre – Vi Agroforestry Programme
SIDA	Swedish International Development Agency
SLM	Sustainable Land Management
SLP OSD	Sierra Leone Police's Operational Support Division
SPGS	Small Production Grant Scheme
SUB	Sustainable Use of Biomass
TANAPA	Tanzania National Parks Authority
tCO$_2$e	Ton of carbon dioxide equivalent
ULA	Uganda Land Alliance
UNDP	United Nations Development Programme
UNEP	United Nations Environment Programme
UNFCCC	United Nations Framework Convention on Climate Change
UN-REDD	United Nations Collaborative Programme on Reduced Emissions from Deforestation and Degradation
USAID	US Agency for International Development
VCM	Voluntary Carbon Market
VCS	Verified Carbon Standard
VCU	Verified Carbon Unit
VPO-Environment	Vice President-Department of Environment
WAPFoR	Western Area Peninsula Forest (Sierra Leone)
WB	World Bank
WHH	Welthungerhilfe
WNHS	World Natural Heritage Site
WWRPTF	Wildlife Works REDD+ Project Trust Fund
ZAWA	Zambia Wildlife Authority
ZDC	Zone Development Committee

AUTHOR BIOGRAPHIES

Albert Arhin is currently a Gates Scholar and a PhD candidate at the Department of Geography, University of Cambridge. A development policy planner by original training, Albert has an interest in interdisciplinary research that uses theories from both the natural and social sciences to understand the travelling policies and discourses, politics and the complexity surrounding natural resource governance and poverty reduction interventions. His PhD research focuses on REDD+ policy processes in Ghana and the diverse pathways for (not) achieving transformational change in the forestry sector. He was born, bred and raised in Ghana, West Africa.

Joanes Atela holds a PhD in Environment and Development from the University of Leeds, UK. He has over five years' research experience in the area of natural resource management, agriculture and rural development. His current research focuses on multi-level environmental policy analysis and implementation for sustainable development.

Vupenyu Dzingirai is a graduate of the University of Zimbabwe. He is an Associate Professor at the Centre for Applied Social Sciences, University of Zimbabwe, where he teaches Research Methodology and Environmental Communication. Also a trained journalist, Professor Dzingirai has carried out extensive research on conservation and development and has done fieldwork in the Zambezi Valley – his doctoral site. His PhD was on human movements and natural resources management in Zimbabwe. Professor Dzingirai resides in Harare.

Ishmael Hashmiu grew up helping his parents with farm work in Ghana and by so doing developed a strong connection with nature. This inspired him to pursue a BSc in Natural Resources Management at KNUST in Ghana in 2001. Thereafter, he played pivotal roles in Sacred Grove Conservation Projects at GACON in

Ghana, as well as in Environmental Restoration Projects at EarthCorps in the USA. In 2008, Ishmael pursued an MSc in Sustainable Environmental Management at the University of Greenwich, UK, where he developed much interest in climate change issues. In 2011 he received a Future Agricultures Consortium Early Career Fellowship to study carbon forestry and livelihoods.

Martin Kijazi studies impacts of global and national environmental governance regimes on local communities' economic welfare and democratic representation. With his IDS-funded research, Dr. Kijazi expanded on his current research of global and national climate change mitigation and adaptation policies of donors, NGOs and national governments. Particularly, how such interventions affect local people's representation and livelihoods. Hence, how they can be tailored better to achieve outcomes that are also representative of local people's needs and aspirations. He obtained his PhD from the University of Toronto, where he focused on forest resource economics and the political economy of development.

Misael Kokwe is a farming systems research agronomist, currently coordinating the Climate Smart Agriculture Project activities at FAO's Zambia office. Misael's research concerns the interface between natural resources management, policy implications and sustainable rural livelihoods in Zambia and Southern Africa. He is currently involved in the development of the REDD+ strategy and implementation framework for Zambia.

Melissa Leach is the Director of the Institute of Development Studies (IDS) at the University of Sussex. She directed the STEPS Centre from 2006 to 2014 and is co-Chair of the Science Committee of Future Earth. A social anthropologist and geographer, her research in Africa and beyond has integrated social science with science-policy and natural sciences across environmental, agricultural, health, technology and gender issues, with a longstanding focus on West African forest landscapes. Her books include: *Misreading the African Landscape*; *Reframing Deforestation*; *Dynamic Sustainabilities: Technology, Environment, Social Justice*; and *Green Grabbing: A New Appropriation of Nature?*

Lindiwe Mangwanya is a doctoral candidate at the University of Zimbabwe in the Centre for Applied Social Sciences. She obtained her postgraduate degree in Sociology at the University of Zimbabwe in the 1990s and another Masters degree in Social Ecology from the same university. Her research focuses on natural resource competition, and has mainly concentrated in the Hurungwe of Zimbabwe where – for her PhD – she examines land struggles in the aftermath of tsetse eradication.

Guni Mickels-Kokwe is a social scientist, researcher and consultant. She has worked in several rural development projects in Zambia over the past 20 years. Her past research examines the linkages between food security, natural resource use and management, rural livelihoods and local empowerment in Zambia. Her current work explores storytelling as a method in constructing historical environmental narratives for action research.

Adrian Nel is an early career researcher with experience researching and teaching around Green Economy, human–environment relations, the political ecology of conservation and development, and responses to crisis of corporate–state capitalism and climate change in the anthropocene. He received his PhD in Geography at the University of Otago in 2014. This concerned the political ecology of carbon forestry and changes to forestry governance under market environmentalism in Uganda. He is from Bulawayo, Zimbabwe, undertook his initial undergraduate and postgraduate studies at Rhodes University in South Africa, and is currently a Visiting Fellow with IDS while working on publications from his thesis.

Ian Scoones is a Professorial Fellow at the Institute of Development Studies (IDS) at the University of Sussex, and is director of the STEPS Centre. He works on the intersections of science and policy, particularly around environment, land and agriculture in Africa. Recent books include: *The Politics of Green Transformations*; *Dynamic Sustainabilities: Technology, Environment, Social Justice*; and *Green Grabbing: A New Appropriation of Nature?*

Thomas Winnebah is a Senior Lecturer in Geography and Development Studies at Njala University and Director of the Environment Protection Agency's Board, Sierra Leone. He has a degree in Geography from Sierra Leone and a higher degree in Rural Sociology from Louisiana State University, USA. He has researched environmental change including in the Gola Forest, Loma Mountains and the Outamba-Kilimi National Park in Sierra Leone and Guinea. He lived and received up to tertiary education in the south, west and central parts of Freetown next to the then WAPFoR's boundary, Sierra Leone.

PREFACE AND ACKNOWLEDGEMENTS

Tackling climate change is one of the most pressing challenges of our age and, as this book is published, the international community is negotiating a new framework for climate governance to be discussed at the Conference of the Parties meeting in Paris in December 2015. This is a crucial moment, and forests, carbon and their management are high on the agenda.

Deforestation and land degradation globally contribute significantly to carbon emissions and addressing these issues has become a major policy priority. Carbon offset approaches, mediated by carbon markets and facilitated by international accords and global climate finance, have become especially popular. In such schemes, carbon emissions in one part of the world (usually the industrialized north) are offset by initiatives that reduce emissions in another part of the world where there are plentiful forests, and opportunities for new carbon sequestration (such as Africa). Such projects can, it is argued, additionally focus on poverty reduction and biodiversity protection, creating a 'win–win' scenario.

This is the theory. But what of the practice? This book is about what happens on the ground when carbon forestry projects – existing in various guises, often under the umbrella of the Reduced Emissions for Deforestation and Forest Degradation (REDD+) programme – arrive. In this new field of environment and development practice, there are many new players, a whole panoply of models, processes and procedures for verification and monitoring, and a hot politics of authority and control. Understanding what works, and what doesn't, is crucial, and this book offers some salutary lessons on the feasibility and desirability of market-based offset approaches to carbon mitigation, and suggests some new directions for the future.

Through a series of detailed case studies from seven African countries, from East, West and Southern Africa, the chapters ask what actually happens when carbon forestry projects unfold in particular places: who wins, who loses out and what are the consequences – for carbon sequestration and offsetting, as well as poverty reduction? As all the cases show, carbon projects do not arrive on a blank slate. All sites have long histories of intervention, including a whole array of forestry, environmental protection and development projects. These have shaped and reshaped livelihoods and landscapes, and generated experiences and memories that influence local responses to new interventions. The book asks: What difference does carbon make? What political and ecological dynamics are unleashed by these new forms of commodified, marketized approaches?

This book emerges from the ESRC STEPS Centre's 'political ecologies of carbon' project, but has been complemented by inputs from others, supported through the Future Agricultures Consortium's Early Career Fellowship Programme, the Responsive Forest Governance Initiative of the University of Illinois, IUCN and CODESRIA, and a PhD undertaken at the University of Otago in New Zealand.

The research team met in December 2013 at IDS, Sussex to discuss the fieldwork, and explore common themes, and these were in turn shared at the 'Green Economy in the South' conference in Tanzania in August 2014. We would like to thank everyone at these events, plus several anonymous reviewers for their comments on the book's themes. We would like to acknowledge Lars Otto Naess and James Fairhead for comments on the introductory chapter. Finally, the book's final stages were expertly overseen by Naomi Vernon, who applied her usual meticulous skills in the copy editing of the full text.

Melissa Leach and Ian Scoones
STEPS Centre, Institute of Development Studies, University of Sussex,
February 2015

1

POLITICAL ECOLOGIES OF CARBON IN AFRICA

Melissa Leach and Ian Scoones

Introduction

The last decade has seen a wave of forest carbon projects across the world, many in Africa. These have been a response to the pressing challenges of climate change mitigation. Conserving or enhancing forest carbon stocks is presented as a way both to reduce carbon emissions from deforestation and, most importantly, to offset emissions elsewhere. A range of new market-based mechanisms have been put in place to facilitate a variety of offset arrangements through payments and trade in carbon credits. This is occurring through a variety of institutional arrangements; some, such as the Clean Development Mechanism (CDM) and the Reduced Emissions from Deforestation and Degradation (UN-REDD and REDD+) process are formally linked with compliance mechanisms associated with international climate change negotiations and the Kyoto Protocol, while others are linked to voluntary carbon markets, regulated in different ways (see Arhin and Atela, this book; Fong Cisneros, 2012). A mass of literature is now asking how forest carbon projects are unfolding, how they might be most effectively geared to climate mitigation challenges and how forest users might benefit from them (e.g. Angelsen *et al.*, 2009, 2012; Corbera and Schroeder, 2010; Blom *et al.*, 2010; Sunderlin *et al.*, 2014a, 2014b; Luttrell *et al.*, 2013; Pokorny *et al.*, 2013; Schroeder and McDermott, 2014), as well as how such initiatives are presented in the media and policy discourse (Di Gregorio *et al.*, 2013). At a larger scale, others have been examining the institutional architectures, funding mechanisms and regulatory and governance challenges of the new carbon economy, and its associated interventions (Karsenty, 2008; Angelsen, 2008, 2013; Vatn and Angelsen, 2009; Boyd and Goodman, 2011; Goodman and Boyd, 2011; Boyd *et al.*, 2011; Lederer, 2012a; Stripple and Bulkeley, 2013). Others have been tracking the effects of volatile carbon prices and the evolution of particular markets, highlighting in recent years the low prices and lack of market spread, as well as irregularities, scams and market politics (Stephan and Paterson, 2012; Peters-Stanley *et al.*, 2013; Lane and Stephan, 2014).[1]

This book contributes to these debates through the exploration of empirical cases from Africa, but goes beyond them in two important ways. First, we situate our analyses within a broader understanding of political-economic processes, and particularly the commoditization of nature and environment. Second, we are interested in landscapes, and how forest carbon projects are interlocking with and reshaping these. Together, these political-ecological dynamics are generating a range of carbon conflicts that are at once both material and discursive. They have profound implications for whether and how forest carbon interventions are designed in the future, and who will gain and lose from them.

Forest carbon policies and projects are part of a more general move to address environmental problems through attaching market values to nature and ecosystems. Under the rubric of the 'green economy' and conserving 'natural capital' (UNEP, 2011), a variety of payment and offset mechanisms are becoming a dominant mode for environmental policy and action (MA, 2005; Bateman *et al.*, 2011), including payments for ecosystem services, and biodiversity and species offsets (Redford and Adams, 2009; Kosoy and Corbera, 2010; Brockington, 2011; Sullivan, 2013). Such commodification of environment and resources aligns with 'neoliberal' economic policies, in a particular phase of capitalism involving a combination of privatization, financialization and appropriation (McAfee, 1999, 2012; Harvey, 2006; Igoe and Brockington, 2007; Castree, 2008a, 2008b; Fletcher, 2010; Büscher and Dressler, 2012; Büscher *et al.*, 2012; Fairhead *et al.*, 2012), and a recasting of the role of the state in environmental management. Forest carbon projects – often to date analysed in isolation – must be seen in this context.

However, they must also be situated as part of lived-in landscapes that intersect with such market-driven processes. In Africa, new carbon-focused interventions are taking place in forest landscapes with particular histories, embedded dynamic ecologies, social and property relations, livelihood practices, knowledge and understanding and above all, politics (Fairhead and Leach, 1998; Ribot *et al.*, 2006). The places now targeted for forest carbon projects also have long histories of external intervention in the name of environment and development, whether focused on colonial timber and mineral extraction, watershed protection, wildlife and biodiversity conservation, agroforestry or modernizing forest-based livelihoods – each justified by particular views of forest use and change (Leach and Mearns, 1996; Fairhead and Leach, 1998). The material and discursive legacies of these past interventions have not gone away. Forest carbon projects do not arrive on a blank slate, but in places that have accreted layers of human-environment interactions, memories and imaginaries, institutions, rights and forms of authority.

By stepping back, and asking somewhat different questions to the burgeoning literature on forest carbon, this book is therefore interested in the political ecology–economy of forest carbon projects in historical context, as part of longer-term landscape change, intervention histories and changing market and valuation processes. It asks: How is carbon commoditization and marketization interlocking with long-term pathways of landscape change and political economy, and so reshaping livelihoods and ecologies? Who are the winners and the losers? What new political and ecological dynamics are emerging as forests are revalued for carbon? Or put more simply – amidst ongoing pathways of change – what difference does 'carbon' make?

The book explores and illustrates these questions through seven cases from across Africa that differ across a number of axes. They cover different project types: in relation to tree and land ownership arrangements, and whether the focus is on protecting or planting trees. They represent country cases at different points of integration with the UN-REDD process, and thus with different institutional and policy configurations. They cover a range of national political–economic contexts, involving different state–market relationships and patterns of centralization–decentralization, which both affect forms of authority over forests. They also represent different histories of landscape and intervention. In these different contexts, carbon conflicts occur everywhere – but they take different forms.

The case studies are introduced in more depth below. The next section outlines the core argument of the book, and why it matters.

Political ecologies of carbon in forest landscapes

Throughout this book we are interested in carbon as a substance with diverse meanings and consequences, and a social and political life (cf. Appadurai, 1986). Carbon is part of the carbon cycle, with particular physical and biological properties, and it is part of trees situated in lived-in landscapes, with diverse cultural and economic attributes and values. Yet carbon is also able to become a commodity, isolated from its ecological and social matrix and inserted into particular markets, acquiring a very different set of social and political meanings. This book considers the historical and political context within which this is happening, and how different people – from local forest users to project developers, state agents and international policy actors – understand, become involved in and reflect on this process, and stand to gain or lose from it.

The overt purpose and justification of carbon projects is to tackle global climate change. But they also have a far wider range of political–economic effects, associated with particular interests. The social and political life of carbon is deeply enwrapped with these. Thus, carbon is seen by cash-strapped states in Africa as a source of foreign exchange (Arhin and Atela, this book), allowing a new economic value of forest resources to be unlocked and deployed for economic growth under the sovereign control of the state. For donors and non-governmental organizations (NGOs), carbon enables a new round of 'missionary' development activity, as projects aim to rescue local ecosystems and livelihoods now claimed to be suffering from and contributing to the impacts of global change (Dzingirai and Mangwanya, this book). Carbon is also seen as a business opportunity for brokers, traders, consultants, private companies and others. As another element in Africa's commodity-led economic boom, carbon joins land for food and biofuels, minerals and wildlife and ecotourism sites as a potential source of investment. In the post-financial crisis world, with footloose capital looking for favourable returns, investments in Africa, including in carbon markets, are on the rise. This may involve a combination of foreign companies, speculative finance and domestic or regional elite capital.

As with any commodity boom, a so-called 'resource curse' threatens (Auty, 1993; Sachs and Warner, 1995; Humphreys *et al.*, 2007), with opportunities for elite appropriation, rent-seeking and corruption and lack of accountability to citizens. Others have written of the 'aid curse' in similar terms (Djankov *et al.*, 2008; Moyo, 2009). In the case of carbon, the two curses are potentially linked. Aid investments are geared towards, and indeed are critical for, new enterprises that would otherwise be infeasible given the current state of carbon prices and the high costs of project start-up and capacity-building. Indeed, in some circles, REDD initiatives are now talked of not as market-based approaches, but performance-based aid interventions (Angelsen, 2013). The combination of aid and private-sector activity feeds expectations of new forms of commodity and markets, and aligns neatly with a new aid and development rhetoric in which global public goods are delivered through 'public–private partnerships'. Thus, markets are being co-constructed with particular relationships among states, private-sector players and aid agencies and other international players in the contemporary political–economic context. Such interactions create 'friction' (cf. Tsing, 2005), emerging from contested, globalized interactions across differences of power and culture. They also bring together diverse actors in new forms of sometimes 'awkward engagement' that result in resistance, conflict or negotiation, with diverse, contingent outcomes (see Nel, this book).

Policies and projects promoting forest carbon offsetting all assume and depend on the idea of carbon as a commodity: isolated, tamed, priced and exchangeable. The construction and marketization of carbon as a commodity also rests on and requires particular understandings of landscape and landscape history. In particular, assumptions of ongoing, one-way patterns of deforestation and degradation from the past into the future are critical to justifying project intervention, providing the so-called 'baseline scenario' of carbon emissions that would supposedly occur without it. Project documents generally portray forests as once plentiful but now under threat and likely to disappear unless outside agencies intervene. Moreover, they almost universally blame local people and their practices, justifying interventions to modify, exclude, disenfranchise or even criminalize them. Carbon as a commodity is now assumed to offer the route to realizing value, and so protecting and enhancing forests, while generating trade profits for carbon's new project 'owners'. There is, therefore, an interlocking of particular ideas and discourses of forest cover change – and associated blame for it – with political–economic interests in carbon marketization and profits.

These processes of marketization are embedded in particular practices, many of which are peculiar to the burgeoning carbon industry. A highly technical language has developed of baselines, of stock measurement, of additionality, of reference areas, of leakage and so on (Angelsen, 2008; Calmel *et al.*, 2010; Mercer *et al.*, 2011; Lovell and MacKenzie, 2011; Ascui and Lovell, 2011; Olander and Ebeling, 2011). Different mechanisms – REDD+, the CDM, Voluntary Carbon Standard (VCS) and others – have developed their own detailed methodologies, based on a set of broader principles. These provide an accounting framework for translating carbon in forests into measurable credits that can be marketed, and for monitoring and verifying that the potential for climate change mitigation is actually taking place. The detail and complexity of these requirements, along with others designed to track and safeguard projects' social and environmental impacts, are such that a whole consulting and guidebook industry has developed to help project developers to navigate the process. We argue that this almost inevitably encourages project designs that make accounting for carbon easier. This pushes projects in particular directions, deeply affecting how they engage with landscapes (Leach and Scoones, 2013).

Thus, discourses and practices associated with forest carbon construct and, through justifying particular sorts of project, transform landscapes in particular directions. They create pathways, particular trajectories of intervention and change (Leach *et al.*, 2010). They restructure ecologies, livelihoods and relationships between people, land and resources, and so property relations and institutions. The effects can amount to, and be interpreted in terms of, the phenomenon of 'green grabbing', whereby, at a particular moment in capitalist development, nature and resources become appropriated in a process of accumulation by dispossession, with carbon, once part of peoples' lived-in landscapes, becoming financialized and part of international markets to the benefit of others (Fairhead *et al.*, 2012; Corson and MacDonald, 2012).

Elucidating this co-construction of carbon imaginaries, political economies and landscape re-working in particular African settings is a key aim of this book. But we are also interested in what these processes exclude. We are interested in the alternative ideas and values that exist around carbon, which are associated with particular people and interests. We are interested in the limits of 'tameability', and so market appropriation. We explore if, as part of dynamic ecologies, carbon might, in fact, be rather more 'unruly' than assumed. And we address the 'unruliness' of local forest users and their ideas, which, grounded in their own logics and histories, may not be so easily captured and controlled. How, we ask, does this double unruliness of carbon and people manifest itself in processes of dissonance and resistance to the assumptions and actions of forest carbon projects?

Equally, we are interested in how these processes play out in particular local settings. Here, an ethnographic focus reveals not a one-way view, but a more complex and variegated picture of winners and losers. The book's chapters examine how carbon forest projects interact with forest-based livelihoods, whether around timber production, farming, forest product harvesting and so on, and how different social groups are involved, whether women or men, youth or elders, long-term residents or immigrants, or members of occupational and ethnic groups. Tracking these interactions reveals the importance of local institutions around tenure, property, labour and authority in mediating access and control over resources, and opportunities to benefit or not from project interventions (Cotula and Mayers, 2009; Bond *et al.*, 2009; Larson *et al.*, 2013; Dokken *et al.*, 2014; Naughton-Treves *et al.*, 2014; Sunderlin *et al.*, 2014a, 2014b). Some people are indeed being dispossessed, but others are gaining in income, property and power; who this is and the alignments involved are sometimes surprising.

And while these interventions, and their political-economic drivers, practices and discourses, have become powerful pathways in African forest landscapes, alternatives continue. People's own discourses and practices – or subaltern pathways – involve different forms of landscape interpretation and use, grounded in different values and ideas about trees and land. These alternatives, we show, co-exist and interact with the discourses and practices of carbon projects: sometimes in overt contestation, resistance and protest, sometimes in quieter subversion, and sometimes in more tacit continuance of ways of life and living with forests, even as carbon commoditization proceeds apace.

It is these iterative intersections between politics and ecology, whereby politics shapes and reshapes ecology and vice versa, that form this book's analytical focus. These intersections occur across scales, from the local to the global, shaped by institutions and the wider contemporary political economy. Our argument, therefore, fits broadly within the wide-ranging field of political ecology (Blaikie and Brookfield, 1987; Forsyth, 2003; Peet *et al.*, 2010; Robbins, 2012). Our emphasis on landscape and its discursive as well as material dimensions gives the analysis a special focus on knowledge and practice, whether in relation to local forest use or the development and implementation of projects and policies. The notion of pathways (Leach *et al.*, 2010), as developed through the past and extending into the future, gives our analysis a historical dimension.

The chapters in the book reveal a range of carbon conflicts. Some involve struggles over resources and property, as trees, land and rights and control over them come to be contested in new ways. Some relate to timeframes, with disconnects between project requirements to lock up carbon for the long-term, and local priorities for shorter-term livelihood flexibility and adaptability. Others relate to imaginaries and visions of what carbon is, what projects are really doing and of what landscape uses and futures are desirable. Carbon conflicts are, therefore, both material and discursive.

Despite the huge industry, rhetoric, claims and well-meaning efforts around 'pro-poor' or people-focused forest carbon, the book shows that there is little prospect of this becoming a reality unless these carbon conflicts are addressed. This, in turn, would require a radical overhaul of how carbon projects are conceived, designed and implemented. Drawing from the book's analysis, at the end of this opening chapter we identify a number of possible future scenarios and policy options for carbon forestry in Africa and beyond.

The following sections take elements of these arguments in turn, clarifying key concepts in relation to aspects of literatures on carbon, the environment and beyond, and introducing how the book's case studies speak to these themes. To begin with, we introduce the different cases from across Africa.

Introducing the cases

Following a chapter that provides an overview of the contemporary policy scene for carbon forestry in Africa and more broadly, the book explores seven cases. These are located in seven countries across East, West and Southern Africa. Each has particular characteristics in terms of ecological setting, project development and the roles of different actors; status in terms of market involvement and certification, and plans for benefit sharing and community development; and overall project type. These basic project features are summarized in Table 1.1.

A central set of distinctions in Table 1.1 relates to project type, in the sense of how – and where – carbon is to be sequestered or offset. The cases cover a range of categories which we label as follows. The first is 'fortress carbon', where carbon and its ongoing sequestration are to be protected in existing forests, with clear boundaries to exclude other activities. The Sierra Leone Western Area Peninsula Forest (WAPFoR) project – focused on protecting a long-established dense humid forest reserve now re-valued for carbon – is a clear example of this approach, as is the Tanzania Kilimanjaro project. In a second category, 'ranching carbon', the focus is also the protection of carbon in existing trees in demarcated land areas, but alongside other resource use such as that of wildlife. The dry forest carbon 'ranches' of the Kenya Kasigau project and Zambia LZRP project offer examples here. A third category involves 'farming carbon' where trees are planted anew – like a crop, albeit one that requires long-term protection. This may be on individual farms – as in the Kenya KACP project, which additionally invokes soil carbon – or in plantations – as in the Uganda case. The Ghana case involves both individual farm and plantation planting. A final category we term 'protected tree carbon'. The Kariba REDD project in Zimbabwe exemplifies this approach, where the valued carbon is sequestered and stored by protecting individual trees, but amidst multi-use landscapes in which people live and work.

Although there is some overlap, these are significant categories in relation to the arguments in this book. By involving different ways of storing carbon, they present different technical challenges and methodological implications for measurement, monitoring and verification. More importantly, the categories imply quite different relationships between carbon, trees, people, rights and landscapes. They therefore also make relevant different opportunities and challenges in taming and commodifying carbon, and in negotiating access and control arrangements. And, as we shall see, unruliness – in ecologies and communities – therefore makes itself manifest to different extents and in different ways in these different intervention types.

The cases also highlight other important contrasts. They are sited across a range of places with different ecosystem characteristics and dynamics. At the most humid end of the spectrum, Sierra Leone's WAPFoR involves old-growth, dense humid tropical forest. Moving into the forest-savanna transition zone, the Carbon Credit Project in Ghana is situated in a forest-savanna mosaic where seasonal fire is a major factor in ecological dynamics. Similarly, the Uganda case is in a humid

TABLE 1.1 The case studies – project features

	Tanzania – Kilimanjaro	Kenya – Kenya Agricultural Carbon Project (KACP)	Uganda – Global Woods Kikonda Reforestation Project and New Forests Namwasa A/R CDM project	Kenya – Kasigau	Zambia – Lower Zambezi REDD+ project (LZRP)	Zimbabwe – Kariba REDD	Ghana – Carbon Credit project	Sierra Leone – Western Area Peninsular Forest Reserve (WAPFoR)
Country and project title / location								
Project developer and status	UNDP/GEF/ GoT (Government of Tanzania); public.	KACP, SSC VI Agroforestry; Swedish NGO.	Global Woods and New Forests Company (UK) – Private industrial timber companies.	Wildlife Works; private.	BioCarbon Partners; Sable Transport; BCP; private.	Carbon Green; private company.	Vision 2050; private.	ENFORAC (environmental NGOs forum), Welthungerhilfe (German NGO); government; public/NGO.
Others involved	Compact project (NGO), created by UNDP.	None.	National Forestry Authority, Sustainable Use of Biomass (SUB – local company) for the Kikonda project. The World Bank for the New Forests project.	None.	Musika NGO, Engineers Without Borders, CSEF (Civil Society Environment Fund), UNDP-AMSCO (African Management Services Company), DFID BIF (Business Innovation Facility); EcoPartners LLC, Client Earth.	Environment Africa (NGO); Hurungwe Safaris (private company).	EDC UK company; ERD Ghana Ltd.	EPA-SL, MLCPE.

How is carbon sequestered/offset?	Forest protection and bans on tree cutting on private land; energy switch from biomass to non-biomass.	On-farm agroforestry for tree and soil carbon.	Reforestation of 'degraded' Central Forest Reserve.	Protecting dryland forests.	Protecting forest as buffer to national park.	Protecting existing forests.	Trees grown on farms and in plantations.	Protecting forest in long established forest reserve; planting in buffer zone to protect core sequestration area.
Project type	Fortress carbon.	Farming carbon.	Fortress carbon.	Ranching carbon.	Ranching carbon.	Protected tree carbon.	Farming carbon.	Fortress carbon.
Who owns the land?	State land (central government (park); alienated local government land; customary land/private.	Individual farmers, half with titles, others under customary.	National Forestry Authority (NFA)-owned land that is leased to the companies.	Community land, ranches (private title), Trust land (council).	Game ranch under private title.	Customary, Rural District Council (RDC), leases on safari areas.	Local farmers.	State/government, but claim contested.
Who finances project start-up?	UNDP/GEF, GoT.	World Bank.	Global Woods for the Kikonda project, and International Finance Corporation of the World Bank, HSBC and Agrie-Vie (investment fund) for the New Forests project.	Private international companies.	Combination of private investors, and some donor funds.	South Pole, private investor.	EDC UK and private developer's teak plantations.	EU, NGOs.

TABLE 1.1 continued

	Tanzania – Kilimanjaro	Kenya – Kenya Agricultural Carbon Project (KACP)	Uganda – Global Woods Kikonda Reforestation Project and New Forests Namwasa A/R CDM project	Kenya – Kasigau	Zambia – Lower Zambezi REDD+ project (LZRP)	Zimbabwe – Kariba REDD	Ghana – Carbon Credit project	Sierra Leone – Western Area Peninsular Forest Reserve (WAPFoR)
Which markets?	Developing scheme through UNDP MDG Carbon Initiative, and in future other voluntary markets.	Voluntary markets through the Biocarbon Fund.	Voluntary – Carbon Fix standard, and the A/R CDM.	Voluntary markets through certified emission reduction.	Originally VCS, voluntary markets. Now aiming for REDD + regulated markets.	Voluntary.	Hoping for voluntary carbon plus World Bank Biocarbon fund, eventually nothing.	Hoping for voluntary carbon markets and corporate environmental responsibility deals.
Involvement of state	Central government contributes to funding; regional government implementing the project.	Limited, but initial assessment; extension workers not involved.	Owner of the Central Forest Reserves.	EIA initially, not ongoing.	Wants to be a REDD demonstration project; involvement of state limited.	RDC owner of the land (formally).	None.	Passive.
Certification status and dates	Potential, ongoing; screened positively for UNDP MDG (marketing); use CDM certification.	VCS in 2012.	Carbon – Fix and CCBA for the Kikonda project, 2009; AR–AM0004 ver. 4 for the New Forests project 2011.	VCS and CCBS, 2011.	CCBA from June 2013, VCS ongoing.	VCS and CCBS, 2013.	Nothing.	Nothing yet; hoping for VCS 2014.

Benefit sharing plan	Direct payment of money based on emission reductions to institutions involved; wider population not to receive benefits.	Carbon revenue given as a bonus tied to farmer activities; covers project administration; community benefits unclear.	Approx 90% to the companies, 10% to community; state merely receives revenue from the lease of the land.	One third of money to community, one third to administration, one third to ranch sharehold; revenue from communal land directly to community projects.	Between BCP, Sable and investors; not local communities.	RDC, 30%; Community, 30%; Hurungwe Safaris, 10%; Developer, 30%. Community fund.	£40 per tree over 20 years, with pre-finance; absentee investor.	National Park Trust Fund; mainly aimed at forest protection; spare to civil society projects.
Community engagement and projects	Alternative livelihood projects including grants to community-based organizations.	None at the moment. Plans for community projects.	Employment, failed outgrower scheme, Dam building, Community development project.	Livelihood projects: water, bursaries, classroom construction, job schemes.	Community benefits through projects, and social services.	Micro irrigation, woodlots, beekeeping, anti-poaching, conservation agriculture.	No.	Water management projects; alternative livelihoods projects funded by EU to date.

savanna area. A transition between savanna and dry ecologies where woodland and grassland co-exist in patterns shaped by scanty and variable rainfall provides the setting for the KACP in Western Kenya, while drier vegetation and highly variable rainfall characterize the Kenya Kasigau project, the Kariba REDD project in Zimbabwe and the LZRP project in Zambia. Finally, the Mount Kilimanjaro carbon project in Tanzania covers a mixture of montane and lowland forest.

The cases involve projects developed by different actors and alliances. Several are led by private companies, although in some cases with support also from NGOs or international agencies (e.g. the KACP, Ghana Carbon Credit project, Kariba REDD in Zimbabwe, the Kenya Kasigau, and the Zambian LZRP project). The Uganda case involves private forestry companies leasing government land, assisted with some development finance. In contrast, both the Sierra Leone and Tanzania projects are led by public agencies and NGOs.

The projects are targeting a range of carbon market schemes and opportunities, and are at different stages in involvement or certification in relation to these. Thus, a number of the projects are already certified and selling carbon credits, including the two Kenyan projects, Kariba REDD in Zimbabwe, the Uganda projects and the LZRP project in Zambia. Of the other cases, the Carbon Credit Project in Ghana originally hoped to sell credits to voluntary markets, but in the end applications for certification stalled, and the project itself collapsed with no credits sold, at least for now. Both the Kilimanjaro project in Tanzania and the WAPFoR project in Sierra Leone are developing schemes to seek VCS certification, aimed at targeting voluntary markets and, in the Sierra Leone case, corporate environmental responsibility deals with companies operating in the country.

All the cases claim the intention of community benefit – either through shares of carbon credit sale revenues, or in other forms. But, given the varied stages and statuses of the projects, it is not surprising to find that these plans are also very varied, and in some cases vague and ill-formed. Several of the projects (the KACP and the Ghana Carbon Credit project, for example) planned direct financial benefits to local people, though often at very meagre levels. In contrast, others – such as the Kenya Kasigau and Zimbabwe Kariba REDD projects – have plans that divide carbon revenues amongst community institutions, the project developer and other actors. The other cases envisage no direct financial benefits but instead expect to secure community engagement through funded community projects. Many of these are geared to 'alternative livelihoods' intended also to reduce pressure on the use of forest resources so that the latter can be protected for carbon.

The cases therefore give good coverage of the array of project types, settings and development activities found in the African forest carbon scene today. This range adds to and complements other recent case studies and reviews, whether Africa-focused or broader (Kanninen *et al.*, 2007; Angelsen, 2008; Angelsen *et al.*, 2009; Brown, 2013; Pham *et al.*, 2013; Murdiyarso *et al.*, 2012; Sunderlin *et al.*, 2014a, 2014b). Much of this now large literature is operational in focus, interested in improving projects. Other works take a more critical stance, but are often focused on individual examples (e.g. Corbera and Brown, 2008; Nel and Hill, 2013). By taking a comparative and critical perspective across a set of African cases, we are able to explore in greater depth a number of key themes relating to the discursive, practical and political intersections of landscapes, livelihoods and markets, as we now go on to elaborate.

Landscapes and narratives

We argue in this book that carbon is situated within historically constituted landscapes. But what is meant by 'landscape'? The term has become a buzzword in recent debates about carbon forestry, 'climate smart' agriculture and environmental management more generally, entering the lexicon and policy approaches of major international agencies such as the Center for International Forestry Research (CIFOR, 2014), the International Union for the Conservation of Nature (IUCN, 2014), the United Nations Environment Programme (UNEP, 2014), the World Bank (2014) and the CGIAR Research Programme on Climate Change, Agriculture and Food Security (CCAFS). For these agencies, a landscape conveys an area in which diverse resources and goals are to be managed in an integrated way. Thus for the World Bank:

> A 'landscape approach' means taking both a geographical and socio-economic approach to managing the land, water and forest resources that form the foundation – the natural capital – for meeting our goals of food security and inclusive green growth. By taking into account the interactions between these core elements of natural capital and the ecosystem services they produce, rather than considering them in isolation from one another, we are better able to maximize productivity, improve livelihoods, and reduce negative environmental impacts.
>
> *World Bank, 2014.*

Such a spatially based planning approach is, as in this definition, compatible with a commoditized approach to carbon and other resources and landscape features – all of which constitute elements of 'natural capital' to be managed in an integrated and thus more efficient and cost-effective way. Others see the key role of landscape approaches as providing tools to manage trade-offs – including between environment and development, conservation and livelihoods – through rational spatial planning and the allocation of different areas to different uses. They emphasize their value in integrating and reconciling poverty alleviation goals with those of forest resource protection and enhancement – including for carbon (Sayer *et al.*, 2013; CCAFS, 2013).

In contrast, and drawing on literatures in historical and political ecology, cultural geography and anthropology, we argue for a notion of landscape as simultaneously material, social, historical and discursively produced, representing social imaginaries, cultural inscriptions, social identities and embedded politics

(Williams, 1973; Cosgrove, 1984; Duncan and Ley, 1993; Crumley, 1994; Demeritt, 1994; Mitchell, 1994; Braun and Castree, 1998; Ucko and Layton, 1999). And, as a consequence, rather than simply providing neat rational planning tools, landscapes and understandings of landscape features, priorities and change are often contested. Contests over landscape visions often have long histories; contemporary conflicts over carbon in Africa, for example, may carry the legacies of struggles over domination from the colonial era or before (Luig and van Oppen, 1997; Offen, 2004).

Landscape features and forms are shaped by interactions between people, their social relations and practices, and ecosystem elements and dynamics – in soils, vegetation, water and so on. As political ecology approaches emphasize, such human–ecology interactions, and so landscapes, are shaped by institutions, political economies and struggles over resources, at local, regional and global scales (Blaikie and Brookfield, 1987; Mehta *et al.*, 1999; Ribot and Peluso, 2003; Robbins, 2012). Environmental historians and historical ecologists emphasize the importance of particular historical contexts to such interactions, and the legacy of past relationships for present landscapes (Grove, 1997; Crumley, 1994; Balée, 2006). Historical approaches often reveal landscapes assumed 'pristine' actually to be deeply human-influenced, or anthropogenic (Offen, 2004). This certainly applies to forest landscapes in Africa, where both large forest expanses and patches of woodland once assumed to be 'natural' have been shown to be anthropogenic landscapes influenced by people's past settlement, livelihood and everyday practices, enriching as well as degrading – according to particular social and cultural values (Fairhead and Leach, 1996, 1998). As historical practices and social relations leave legacies on which subsequent practices build, anthropogenic landscapes can be seen as complex '*palimpsests*' embodying the ongoing outcomes of shifting social-ecological interactions over time: 'a vertical or horizontal layering or stratigraphy of [anthropogenic] signatures and patterning etched on the surface of the earth, deep into the soil... or above the surface fixed in the layers of vegetation' (Erickson and Balée, 2006, p.187).

Thus, in Sierra Leone, the WAPFoR reserve (Winnebah and Leach, this book) contains the sites of long-abandoned pre-colonial settlements whose inhabitants' planting of cotton and kola trees is still visible in vegetation patterns in the dense humid forest. In the forest-savanna mosaic of Ghana, where the Carbon Credit Project is located, forest patches often overlie old settlement sites with their enriched soils and historically reduced fire risk (see Fairhead and Leach, 1998; Chouin, 2009). As Kijazi (this book) discusses, humans have continuously inhabited the slopes of Mt. Kilimanjaro for the last 2,000 years (Odner, 1971). Histories of immigration and human settlement have progressively transformed forest into a primary-, secondary-, and agro-forest (*vihamba*) mosaic that some have hailed as a highly sustainable land-use system (Fernandes *et al.*, 1984). Yet current fortress conservation presumes a return to an ideal pristine forest reserve.

Landscapes are also cultural, representing particular ways of seeing, replete with symbols and cultural imagery that interplay with broader beliefs, cultural practices and morals (Cosgrove, 1984). Such ways of seeing often embody ideas about social and moral relations, such that 'landscape constitutes a discourse through which identifiable social groups historically have framed themselves and their relations with both the land and with other human groups' (Cosgrove, 1998, p.ix). While this definition was originally associated with a particular, European idea of 'landscape', anthropologists, cultural geographers and others have since emphasized the co-existence of and contestation between diverse landscape ideas and discourses, including in African forest settings (e.g. Fairhead and Leach, 1996). These may be associated with different local social groups, or with local forest users *vis-à-vis* state, international or policy agencies, who develop perspectives on landscape that reflect their own cultural backgrounds and social positions and interests. Thus, for example, on Mt. Kilimanjaro, the state views the landscape as an important space for development and conservation – settler farm estates on the lower slopes, and protected reserves on the upper slopes – with local people's settlements sandwiched in between. In contrast, local forest users see both settler and conservation estates as alienated ancestral lands.

Particular landscape perspectives are often incorporated into narratives or storylines about landscape change, including ideals of how a landscape ought to be, and ideas of blame, victimhood and heroism for bringing about certain kinds of change (Roe, 1991). Such narratives draw on selective forms of knowledge and theorization of change, whether emanating from formal science or from local experience and people's knowledge. They can be critical in justifying – or contesting – policies and interventions (Keeley and Scoones, 2003). In Africa, environmental narratives incorporating moral judgements about people's assumed destruction of 'natural' landscapes have often been used to justify policies that remove resource control from local users (Leach and Mearns, 1996). Indeed in all the cases discussed in this book, local people are blamed for forest destruction, justifying market incentives or regulatory enforcement to ensure forest protection.

Forest carbon has given new life to longstanding policy narratives about African environments, and about forest landscape change in particular. The widespread storyline about forest decline at the hands of local users found in each of our case studies has been elaborated repeatedly since early colonial times. It draws on scientific and popular ideas and practices that see forest vegetation as 'natural', with people's use constituting disturbance and degradation, only accelerated by population growth (Leach and Mearns, 1996). Portrayals of forest cover and quality as declining, linearly, rapidly and recently, from an earlier state of 'intact' forest, repeatedly construct deforestation as an urgent problem requiring external intervention. Long-established, widely circulating figures suggesting that only 13 per cent of West Africa's 'original' forest cover remains (Sayer *et al.*, 1992) are joined by more recent claims, for instance, that '[a]round the turn of the century, West Africa had some 193,000 sq. miles (500,000 sq. km) of coastal rainforest but today [they] have been largely depleted [...] Now [...] only 22.8 percent of West Africa's moist forests remain, much of this degraded' (Mongabay, 2012, p.1), or that the Guinean Rainforest of West Africa had, by 2000, been reduced to 18 per cent of its original area (Gockowski and Sonwa, 2010). Since early colonial times such convictions of rapid and ongoing forest loss from an 'original' baseline have driven policies to halt deforestation and conserve what are assumed to be remaining forest fragments, whether to safeguard hydrology, agro-ecological productivity, timber or biodiversity 'hotspots' (Bakarr et al., 1999; Conservation International, 2008). Today's new round of carbon-focused initiatives are similarly justified by narratives of rapid forest loss: 'Africa's tropical forests are an important store of carbon [yet] Africa's forests are being lost at around three times the world average' (Mercer et al., 2011, p73).

Forest carbon projects arguably rely even more strongly on deforestation narratives than previous forest policy interventions, since these are necessary to construct a 'baseline scenario' against which sustained or increased carbon stocks can be measured and verified (see below). Thus, in the justifications for each of the case study projects, we find narratives about likely ongoing deforestation in the absence of project intervention. For example, the feasibility study for the Ghana Carbon Credit Project (Hashmiu, this book) claimed that '[r]ates of deforestation in West Africa are among the highest in the world. Ghana's tropical forest cover has decreased from 8 million hectares at the beginning of the 1900s to about 1.6 million hectares in 1990, and the deforestation rate is high: nearly 65,000 hectares per year. Virtually all forest currently left is located in forest reserves' (TREES, 2010, p.10). Commenting that '[t]oday's situation in the project area [...] shows a small-scale patchy structure of different vegetation types' (TREES, 2010, p.22), forest patches were assumed to represent 'remnants' of past extensive forest areas. In Sierra Leone (Winnebah and Leach, this book), the scoping study for the WAPFoR project drew on satellite data to infer a (9 per cent) loss of forest cover in the project area since 2000 (OBF, 2012, p.12). Such recent deforestation fitted logically with widely repeated discourses of one-way deforestation – for instance, that '50 per cent of the country has conditions suitable for tropical rainforest, but less than 5 per cent is still covered with [...] closed forest' (Sayer et al., 1992, p.944). Projecting forward to 2031 by linking this historical deforestation to population growth (OBF, 2012, p.4), the study produced baseline deforestation scenarios that suggested a sufficient level of 'avoided deforestation' to justify the carbon project. In Mt. Kilimanjaro, the GEF/UNDP sustainable land management project (Kijazi, this book) argued that '[t]he ecosystems and watersheds of the Kilimanjaro are experiencing an extensive process of degradation and deforestation' (GEF, 2010, p.2) – attributed to rapid population growth, land use change, poor land management practices, unsustainable harvesting of natural resources and climate change. The proposal used a high baseline deforestation rate of 6 per cent per year, although without explaining the origins of this figure. In the same way, in Uganda (Nel, this book), forests were cast as 'degraded' in order to allow for production zones of the forest reserves to be planted with exotics by companies as part of forest carbon projects, resulting in widespread eviction and displacement of local people. While such narratives contain elements of truth, the assumption of one-way, linear forest loss overlooks both the inappropriateness of asserting an 'original' forest cover in landscapes subject to deep vegetation–climate fluctuations going back centuries (Fairhead and Leach, 1998), and far more complex, multi-way landscape interactions in more recent times.

We also find narratives about local forest–climate relationships – the impact of forest cover and quality on local hydrology, watersheds, weather systems and fire regimes – reinvoked in carbon project justifications. In Ghana, the project was justified as safeguarding micro-scale weather systems offering stability in rainfall patterns and protecting against wildfire risk. In Sierra Leone, the WAPFoR project slogan, 'water in the forest is life', justifies forest protection to preserve the watersheds that provide urban water supplies. Such justifications draw on a mix of (often contested) scientific theory and popular ideas. In carbon projects, such local forest–climate interrelationships are sometimes linked upwards and outwards to global climate change – in imaginative and scientifically dubious ways. Thus, in Tanzania, the melting of glaciers on Mt. Kilimanjaro is portrayed as an effect of local deforestation, while forest watershed protection in Sierra Leone is now discursively linked to protecting forests for global climate change. Given that the relationship between local carbon stocks and global climate is hard to understand and explain – for project staff, policy-makers and publics alike – it is perhaps not surprising that the off-the-shelf narratives about local forest–weather links are inserted into project communications, discussions and reflections in their place – even when they fly in the face of decades of scientific research.

In these ways, African deforestation narratives are now being invoked and indeed strengthened in a new global political and policy context. And they have acquired a new sense of urgency and drama, now interrelated with a global climate crisis and the struggle to tackle it. Similarly, discourses of local blame, extending back to colonial times, are being reinvigorated. These often focus on so-called 'slash and burn' farming – itself a negative description of longstanding and, in many parts of Africa, sustainable bush-fallowing practices (Richards, 1985; Fairhead and Leach, 1998; Palm *et al.*, 2013). In many regions it is no longer practiced, as socio-economic and demographic change has contributed to transformations towards more intensive farming systems. Nevertheless, forest carbon project discourses have strikingly brought 'the slash and burn farmer' back to life, re-imagined as the key villain responsible for forest loss and threat, whose ways external interventions must now seek to amend. In our cases, we find project documents invoking such narratives about slash and burn farming even where it has not existed for decades. For instance, while the World Bank's Country Environment Analysis (CEA) of Ghana blames slash and burn agriculture for the loss of more than 50 per cent of the country's original forest cover (World Bank, 2006), farmers in the forest-savanna transition zone now rarely practise this and, when they do, it is in bush fallows, not old-growth forest (Hashmiu, this book). In Kilimanjaro, various intervention documents make reference to 'poor farming practices' (Kijazi, this book). This is in sharp contrast to evidence from socio-ecological studies (Hemp and Hemp, 2008) that the Chagga home gardens are one of the most sustainable farming systems in the world. In the Kenyan Kasigau project, the LZRP in Zambia and the Kariba REDD cases, for example, the project design documents each emphasize the need for the project to curb the exploitation of the dryland forest through slash and burn practices and small-scale charcoaling by often extremely poor local residents and migrants (see Atela, Chapter 6; Mickels-Kokwe and Kokwe; and Dzingirai and Mangwanya, all this book).

In many of our cases, alternative, marginalized or even hidden narratives, including those of local forest users, tell quite different stories about landscape change. These draw on different forms of understanding, embedded in the material practices and political ecologies through which people live with and shape landscapes in the contexts of their histories and livelihoods. Thus, forest patches in mosaic landscapes are, according to local perspectives and ecological and historical evidence, often not remnants at all but the outcome of vegetation enrichment in grassland, linked to settlement and everyday practices and their effects on soil, fire and vegetation (Fairhead and Leach, 1998; Chouin, 2009; see Hashmiu, this book, on Ghana). Local landscape narratives often include such possibilities of human enrichment. They also invoke a sense of context and site specificity, and of cycles, variability, diversity, and non-linear dynamics, rather than a one-way decline of forest cover. Such cyclical interactions between ecologies and land and vegetation use, influenced by diverse drivers, provide dynamic contexts into which external project interventions slot. Thus, for example, in the Carbon Credit Project area in Ghana (Hashmiu, this book), there was a dominance of savanna grassland early in the twentieth century. By 1983, the area was heavily forested, with cocoa interspersed with large shade trees. A dry period and a build-up of grass in savanna patches then enabled a massive forest fire, radically transforming the landscape back to a much more open form, with 'forest islands' associated with past settlement sites and sacred areas. These open areas became dominated by maize in the context of relatively high prices and market opportunities, and because cocoa was vulnerable to savanna fires. Yet tree planting for carbon has required fire protection, enabling also a re-investment in cocoa. Carbon trees also help to provide shade for the older cocoa varieties that require it. However, newer varieties do not need shade, threatening the future of this carbon–cocoa intercropping system.

As this example illustrates, local landscape processes also encompass inherent unruliness in ecologies, trees and therefore carbon. The dynamics of fire, soils and vegetation are often non-linear and unpredictable. Local practices have often co-developed with such dynamics and indeed are attuned to making livelihoods amidst them. This involves knowledge, but also adaptability and flexibility – to 'hitch a ride' on nature (Richards, 1985) and live with uncertainty (Scoones, 1995) – and to shift activities, practices and uses to suit changing ecological – and political-economic – opportunities and constraints. Such adaptability in the decision-making of local land users, often required from season to season or over timescales of a few years, contrasts strongly with the requirements of forest carbon projects. These are very dependent on long timeframes, with projects seeking to protect a given area of forest land, or maintain trees planted for their carbon, over timescales of 20–30 years, to meet carbon verification requirements. The idea of a 'project' that lasts 30 years is itself strange – carbon projects do indeed take highly projectized forms, yet where else in rural development would one find a project expected to sustain itself for so long (Mickels-Kokwe and Kokwe, this book)?

For local users, a static, stock-based land use over such timescales is alien both to ongoing farming and livelihood decision-making, and to social and property relations. There is a sense of 30 years being a very, very long time away – beyond the lifespan of many current farmers – thus raising challenging issues around generation and inheritance. The Kenyan KACP case is an example here (Atela, Chapter 4, this book). The project is expected to unfold over 20 years, during which farmers must plant and care for trees on their farms. Yet most are women aged between 40–60 years, and unlikely to benefit directly from the project at completion. Such intergenerational uncertainty is complicated by the fact that the youth who are expected to take over project engagement are interested in more immediate returns, and often have a very different relationship to the land compared with their parents.

In the histories of Africa's forest landscapes, current carbon projects are often only the latest in a series of interventions. Thus, for example, in Zimbabwe, Kariba REDD is happening in a setting that has experienced multiple displacements and population movements, from those linked to colonial settlement of white farmers, to displacements due to the Kariba dam construction and the establishment of national parks, to more recent movements due to land reform (Dzingirai and Mangwanya, this book). WAPFoR in Sierra Leone has been a forest reserve since 1916, with successive waves of production and protection interest focused on timber, watershed protection, biodiversity and now carbon (Winnebah and Leach, this book). In the Mt. Kilimanjaro case (Kijazi, this book), the recent carbon interventions are happening in a landscape with a long history of land alienation for settler estate farming and the creation of protected areas. Most recently, in 2005, the national park was massively expanded, and this was followed by a tree-cutting ban on all private and public lands in the region. In historical ecology terms, forest landscapes can therefore be read not just as layered human–ecological interactions, but also as layered external interventions. As the cases reveal, the legacies of past interventions in terms of ecologies, tenure relationships and people's attitudes to the state, amongst other factors, in turn influence how more recent interventions focused on carbon unfold. At the same time, these projects are not just 'more interventions'; the carbon focus also makes some important differences.

Carbon constructions and imaginaries

What, then, is the meaning of carbon in the context of such interventions? How is carbon defined and understood? How is it imagined, viewed and constructed as part of different narratives, and in relation to particular practices – those of projects, and those embedded in landscapes and livelihoods in alternative ways?

The notion of 'imaginaries', as explored in cultural studies, is helpful here. Social imaginaries refer to collective visions of attainable futures (Taylor, 2004);

they are at once descriptive of how things might unfold, but also prescriptive, suggesting guides for action or policy (Harvard, 2012). More specifically, socio-technical imaginaries can be defined as 'imagined forms of social life and social order that center on the development or fulfilment of innovative scientific and/or technological projects' (Jasanoff et al., 2007, p.1). Interventions to value carbon and link its trade to the mitigation of climate change can be seen as just this kind of innovative endeavour, in turn requiring specific socio-technical imaginaries of carbon to suit them.

The core imaginary of carbon, co-constructed with the field-level initiatives explored in this book, is as a commodity. Thus, '[m]arkets in greenhouse gas emissions are organized around carbon dioxide equivalence to create "exchange value" and a fungible commodity that can be traded across products and projects' (Newell et al., 2012, p.4). What is required for a market in carbon credits to function is commensurability (Lohmann, 2009), so that a tonne of carbon conserved in a forest is equivalent to that emitted, for instance, from an industrial factory in Europe or the USA. This is not straightforward; it requires extracting and isolating carbon, conceptually and materially, from the territories, histories, economies and politics in which it is embedded, and from dynamic cycles that involve oceans and atmosphere as well as vegetation, so that it can be re-conceptualized and financialized as tradeable units. The concept of 'tonne of carbon dioxide equivalent' (tCO_2e) widely used in climate change market and policy literature captures this idea of equivalence. As Bumpus (2012, p.17) points out, 'carbon offsets create a commodity and value out of a piece of nature – carbon dioxide in the atmosphere – that if achieved properly, *does not exist*' (emphasis in original). This is because several 'types' of carbon in an offset project are supposed to cancel each other out: the carbon that continues to be emitted by the offset credit buyer, and the carbon that would have been emitted if it had not been displaced by the project activity (e.g. carbon in trees associated with avoided deforestation). These 'types' of carbon exist materially in different places, forms and conditions. Carbon markets rely on constructing them as equivalent, and so exchangeable.

For forest carbon projects to render the carbon that cycles through forests into exchangeable tCO_2e involves a series of discursive and practical moves. A first involves abstracting the trees that constitute 'carbon stocks', and the sequestration processes that reduce atmospheric carbon, from wider landscapes and their dynamic ecologies. Second, carbon needs to be re-conceptualized as a 'unit' of nature that is amenable to exchange. Castree (2008b, p. 280) terms this process of categorizing and separating out a thing from its supporting context 'individuation'. Subsequently, these individuated units need to be financialized – re-imagined and valued in monetary terms. Financialization is the process of drawing into financial circulation aspects of life that previously lay outside it; of attempting to reduce all value that is exchanged (whether tangible, intangible, future or present) into a financial instrument. Financialization has been identified as a critical precondition for the emergence and operation of diverse offsets and markets for 'nature' (Igoe et al., 2010; Büscher, 2011; Sullivan, 2011).

Abstraction, individuation and financialization are thus three essential processes involved in the commoditization and sale of carbon. Yet carbon commoditization is distinct from the commoditization of bits of nature for many other purposes, such as a tree for timber (Bumpus, 2012). Whereas timber units retain essentially the same materiality throughout the commoditization process, carbon units need to be (re)imagined as equivalent to emissions reductions in distant places. This gives carbon commoditization a peculiar character, interlinking it with the quite novel imaginaries associated with the new global carbon economy (Bridge, 2011; Goodman and Boyd, 2011; Newell *et al.*, 2012; Knox-Hayes, 2013; Kama, 2014) and contemporary forms of neoliberal nature (Büscher *et al.*, 2012).

The same 'bits' of nature are also part of quite different imaginaries. Thus, the trees, soils, vegetation and atmospheric processes now being imagined, valued and commoditized in terms of carbon are simultaneously embedded in lived-in landscapes whose inhabitants value them in quite different ways, according to diverse socio-cultural perspectives and collective visions of diverse futures. Thus, the same group of trees might be valued by women as an important source of gathered products, essential for the current and future food security of their households; by elders and others as markers of historic settlement, places of social memory and ancestral worship important to securing future community prosperity; by entrepreneurial youth as potential timber resources that could be sold for economic value; or by others as providing cool shady places where spirits reside. These distinct ways of valuing are not commensurable with each other, and indeed may be the subject of ongoing negotiation and debate in local social and political life.

The case studies reveal many examples of such local cultural valuations and negotiations. For example, the Kenyan Kasigau case reflects a diversity of perspectives differentiated by wealth (Atela, Chapter 6, this book). Poorer community members value the dryland forest for the immediate goods of charcoal and firewood, expecting the project to compensate them for the loss of this use value. Wealthier households, by contrast, value forests in terms of recreational and environmental service benefits, as well as highlighting the importance of forest shrines on the hills.

Crucially, though, none of these diverse expressions of value are compatible with those of carbon simply as an exchangeable commodity. Hence, as forest carbon projects emerge onto the scene, bringing with them the peculiar carbon imaginaries outlined above, mutual incomprehension and tension is almost inevitable. On the one hand, the complex and peculiar manoeuvres of carbon commoditization and their links with carbon credit markets are difficult to understand, let alone explain simply – something that can challenge even project developers, and certainly their outreach workers. On the other hand, as projects arrive in communities they meet both very different imaginaries of landscape and ecosystems, and the legacies of communities' past experiences of external interventions, and the ideas conditioned by these. Creative responses result.

For instance, in attempting to describe and pin down what and where 'carbon' is, villagers in Tanzania identify it with 'charcoal air' (*hewa ukaa* or *gesi ukaa* in Kiswahili), resonating with recent climate mitigation campaigns. People from Badu and Dumasua villages in Ghana were told it was 'the smoke you see from an aeroplane, to be absorbed by trees and turned into cash' as part of project promotion. Villagers in Sierra Leone associated carbon with 'the mists you see above forests in the morning', but also with 'the trees growing in the forests, that foreigners want to harvest for money'. The notion of carbon as a kind of extractive resource, that foreigners seek to discover and take away to sell, rather like a mineral, surfaced frequently in both the Sierra Leone and Zambia cases, reflecting, perhaps, people's long experience of foreign mineral extraction in these countries' political economies and histories. Carbon, in this sense, is just the latest in the long line of bits of nature that foreigners have come to extract and sell. In a similar vein, others associate carbon directly with the people – usually foreigners or elites – interested in it. In Zambia, project promoters are referred to as *BaCarbon* – the carbon people. In Sierra Leone, villagers asked researchers – initially presuming them linked with the rumoured carbon project – whether they were 'Reg (aka REDD) who has come to buy our carbon'.

In some cases, project outreach workers have taken up and elaborated on such local narratives in attempts to make carbon projects more locally legible, and to encourage project acceptance by communicating project messages in terms that might have local traction. Thus, in the WAPFoR project in Sierra Leone, project workers built on local ideas about carbon as 'smoke' in messages that the carbon project would 'clean up smoke' globally. While such attempts might be applauded as instances of local extension workers' innovative cultural brokering (Lewis and Mosse, 2006), they are also replete with instances of mis-translation and the creation of further ambiguity and confusion. Moreover, there is a fine line between explanation and education, persuasion and coercion. The case studies provide several examples of carbon projects' so-called 'awareness campaigns' that became, in practice, geared to disciplining and control.

What all the cases illustrate, in different ways, are the major challenges of bridging global carbon commoditization imaginaries, with local imaginaries embedded in landscapes and livelihoods. These make carbon projects even more difficult to incorporate into local settings than many of their predecessor interventions in forest-farmland areas. Carbon projects easily become further examples of alien interventions introduced by outsiders who bring peculiar ideas – as have many prior interventions in wildlife conservation, agroforestry, or rural development before. Such earlier interventions also brought initially alien ideas – the notion of 'biodiversity', for instance – but these more readily found traction and translatability into local languages and conceptual framings. Carbon projects come with a more deeply alien and largely unfathomable logic, which, when struggles to accommodate fail, are easily dismissed as the bizarre ideas of ignorant, though clearly profit-seeking, outsiders.

Generating value: Techniques, measurements and their consequences

These relationships between carbon and its commoditization, and narratives about landscape change, are, in turn, deeply interlocked with the practices and techniques of measurement in carbon projects, as 'the creation of exchangeable tCO_2e relies on the implementation of project activities and the processes of calculating, justifying and verifying emissions reductions' (Bumpus, 2012, p.16). Tensions have to be navigated between the materiality of the carbon and the real-world contexts in which it is reduced or sequestered from the atmosphere, and the institutional requirement, set by carbon standards, to assert that a reduction has taken place against the baseline (non-intervention) scenario. This requires what Bumpus (2012, p.20) describes as a 'hemming in' of carbon dynamics, achieved through a range of practices and methodologies for measurement, monitoring and verification.

Numerous measurement and monitoring, reporting and verification (MRV) procedures and protocols have been developed to legitimate the production and sale of carbon. These are associated variously with the REDD+ process, CDM and VCS standards, and with a range of carbon project types. Our case study project types all fall within what the CDM calls 'agriculture, forestry and other land use' (AFOLU) approaches. AFOLU project categories and associated methodologies include 'Afforestation and Reforestation' (A/R) under CDM, or the VCS equivalent 'Afforestation, Reforestation and Revegetation' (ARR). This involves planting trees or otherwise converting non-forest to forest land, or increasing carbon stocks in woody vegetation (CDM, 2013). By contrast, REDD-type projects involve avoiding 'unplanned' conversion of forests to non-forest areas (deforestation), or reduction of carbon stocks (degradation) (VCS, 2013b). If the project involves avoiding otherwise planned logging or farming, it counts instead as Improved Forest Management (IFM) or Agricultural Land Management (ALM). These categories cover the range of project types we have identified in our case studies, although, as discussed above, the cases vary in which, if any, standards and therefore measurement protocols they have sought to use.

AFOLU approaches have been developed by a particular constellation of climate modelling, environmental economics, biological, accountancy and project management expertises – as represented on the AFOLU committees for both the CDM and the VCS advisory groups (VCS, 2013c). The associated measurement and verification methodologies and practices can be understood as having a social and political life that co-develops with that of carbon-valued-as-a-commodity (Leach and Scoones, 2013). Measurement and modelling of carbon – as sociologists of science have observed for other fields of modelling – is a social process that incorporates and affirms certain social, political and moral assumptions, while excluding others (Morgan and Morrison, 1999; Magnani and Nersessian, 2009; Morgan, 2009).

While approved methodologies vary in detail, all share a set of basic elements (see also Arhin and Atela, this book): *demarcating the project boundaries* and their spatial extent; *ensuring land eligibility* – in relation to vegetation and tenure; *establishing a baseline* – including a change scenario in the absence of project activity, and a reference area; *demonstrating additionality* – providing assurance that the claimed carbon effects would not have happened without the project; *quantifying carbon emission reductions* through new project activities; *assessing leakage* that might occur through displacement of activities from the project site; and *evaluating non-permanence* – assessing the risk that the project's carbon sequestration effects will not last. These generic methodological elements themselves carry with them particular assumptions about forests and landscapes, and carbon as a commodity. Equally, their application necessarily relies on particular practices in collecting and interpreting data. As the case studies demonstrate, methodological protocols and practices thus help to shape and affirm certain landscape narratives and potential pathways of change, while excluding alternatives.

Thus, for instance, we have already seen how the construction of baseline scenarios in both the Sierra Leone WAPFoR project and in the Ghana Carbon Credit Project reinforced longstanding narratives about linear deforestation. 'Additionality' assessments contributed further. In the Ghana case, the project argued for additionality by deploying the standard narrative around deforestation in the area, evoking an image of a past pristine forest being converted to savanna, especially by 'slash and burn' agriculture. Any intervention to protect so-called 'remnant' forest tracts or to plant trees to replace assumed lost forest is thus seen to reverse the trend. This was despite the feasibility study's satellite analysis which was more uncertain, finding that 'the natural forest and the teak plantations within the project area are not identifiable, and the project area vegetation cannot be distinguished from the outside project vegetation. This leaves room for some interpretations which would negatively impact the feasibility of a carbon project' (TREES, 2010, p.23). The study also questioned whether project activities were really distinct from the 'common practice' of tree planting, community forestry and agroforestry projects in the area, dating back over decades (Hashmiu, this book).

In Sierra Leone, the scoping study argued for additionality on the grounds that:

> [T]he WAPFoR is currently under severe pressure, especially from rapid urban expansion/encroachment into the reserve. The business as usual scenario is characterized by low levels of law enforcement, little staff capacity, little human resources, little financial means for effective protected areas management. Consequently, there will be limited means to mitigate emissions without the project.
>
> *OBF, 2012, p.8.*

Again, this argument reworks long-established narratives about ongoing deforestation problems that can be 'solved' only by external intervention – in this case the imposition of strengthened forest protection mechanisms. In both these cases, practices for measuring and accounting for carbon stocks, assessing leakage and evaluating non-permanence pushed the projects towards focusing on (or imagining landscapes as) static, stable and easily measurable, ignoring more uncertain dynamics.

Such disciplining effects of 'hemming in' by project practices are not confined to 'top-down' approaches; they can equally happen when measurement, monitoring and verification is conducted through 'community' assessments using so-called participatory methods. Indeed such participation is required by the Climate, Community and Biodiversity (CCB) standard, and was attempted in the Kenyan KACP case where farmers were expected to complete detailed farm-level records about farm management practices, feeding into more advanced technical accounting procedures. In the Zambia LZRP case, NGO-facilitated workshops provided data on community involvement in forest use which fed into the project design. Yet, despite participatory rhetoric, the assumptions of the measurement protocols tend to prevail and, in turn, to reinforce certain types of project. Thus, methodologies helped to push the WAPFoR project in Sierra Leone and the Mt. Kilimanjaro project in Tanzania towards their 'fortress carbon' approach, while in Ghana what might have been a 'tree protection carbon' or more flexible mosaic landscape approach became instead an easily measurable and controllable 'plantation carbon' project.

Regardless of their particular assumptions, the very multiplicity and complexity of standards, measurements and MRV techniques can be bewildering. CDM, VCS and other voluntary and private approaches are constantly multiplying, with methodologies continuously evolving and being updated. The respective websites have plenty of documents and guidance sheets to download, but these are not for the fainthearted, given their length, multiplicity and sometimes obscure terminology and technical requirements. In response, NGOs and consultancy firms have produced a large array of guides and manuals to help project developers navigate these challenges (e.g. Ingram *et al.*, 2009; Pearson *et al.*, 2009; Calmel *et al.*, 2010), while opportunities have blossomed for consultants to conduct project development operations. Indeed, consultants have been involved at various stages in all our case study projects. The co-existence and layering of multiple private, official and voluntary accreditation processes, each with their own standards, methodologies, application forms and procedures, consultants and brokers, in turn adds to the institutional complexity, ambiguity and competition over carbon project development. New and specialist sources of expertise are required – in measuring, accounting, GIS mapping, modelling and so on. While in some cases this has provided new learning and employment opportunities for local and national researchers and consultants, very often the simplest solution for projects has been to turn to the burgeoning international groups offering such services, in turn supporting the growth of this part of the new global carbon political economy.

In this context, the practical and funding difficulties of carbon project development in African resource-poor settings are very high. Indeed, in some of our case studies, these have proved insurmountable: the Carbon Credit Project in Ghana failed to seek formal accreditation, partly because of the complexity of the process, while in other projects accreditation processes have led to long delays (Hashmiu, this book). The need for donor funding to contribute to start-up costs reinforces the reliance of forest carbon projects on external actors. This, in turn, reinforces the tendency for carbon projects to be seen as not locally owned, but foreign.

Furthermore, in some of our cases, project developers have explicitly claimed the complicated technical requirements as a justification for excluding local communities. In both the Tanzanian and Zambian cases, technical complexity was seen as too difficult for local communities to understand and engage in, so legitimizing their non-participation. In the Zambian project, for instance, donors resisted local plans to develop decentralized GIS capabilities. Control over measurement techniques and practices thus supports control over project directions by the global carbon industry that stands to benefit from them. This further undermines democracy and accountability in forest carbon projects, reinforcing their contribution to local disenfranchisement.

Creating markets

Generating value from carbon means creating markets. But this is not straightforward. Markets are constructed through complex socio-technical processes; they are embedded in social relations and governed by politics (de Alcántara, 1993; White, 1993; Guyer, 2004, 2009; MacKenzie et al., 2007; Mitchell, 2007; Böhm and Dabhi, 2009). Carbon markets, just as any other, are not just the result of supply, demand and resulting prices; their rules and operation are actively created by a range of players and practices (MacKenzie, 2009, 2010; Callon, 2009; Lederer, 2012b).

As the previous section showed, a range of measures and metrics is used to establish carbon as a commodity that can be traded. These interact with accounting mechanisms and practices to give carbon value in a market (Lovell and MacKenzie, 2011; Lansing, 2012). A carbon market must operate in ways that all market actors can comprehend. This requires a set of strategic simplifications and boundary definitions. In a process of containment and 'taming' (cf. Çalışkan and Callon 2009, 2010), carbon that was unruly and dynamic in its original context becomes controlled, auditable and tradeable. This is how value is created, and can in turn be appropriated.

How the market is constructed in turn defines who can participate and on what terms. At each step there are processes of inclusion and exclusion. Much of this is mediated by particular forms of expertise, as many people become enrolled in market construction, along with the tamed commodities concerned. Markets are constituted through the coming together of different actors in different configurations: financiers, project developers, consultants, brokers, guarantors, aggregators, regulators and more. These market players each have different interests, and must appropriate value from their engagement. In the complex world of carbon markets there are multiple players spread across the world. Only some can meet the standards required, only some can comply with the accounting and audit requirements, only some can therefore appropriate value. As a market becomes more elaborate and more spatially dispersed it becomes increasingly removed from local contexts, and the possibilities of local forest users – notionally the 'owners' of the carbon – to become involved is progressively diminished. These market networks are governed by power relations that influence the possibility of negotiation. While 'benefit sharing' protocols may be included in the project specifications and may be a requirement of the standards, the ability to demand a share is affected by the contours of power that construct the market. NGOs and others have pushed strongly for 'safeguards' to provide standards for equity and transparency in market operations, but despite their incorporation into key formal frameworks (such as the Warsaw Framework for REDD), adherence and implementation again depends strongly on power relations, and is often weak.

Again, these socio-technical and political processes of market formation are not peculiar to carbon markets. However, as a novel market around a commodity that only has value in the context of a complex offsetting arrangement operating at a global level, carbon markets are perhaps especially complex, and subject to flux and negotiation (Spash, 2010; Lansing, 2012).

Since their emergence, carbon markets have had a rocky ride. Initially seen as a speculative opportunity, they attracted plenty of attention, including from those with little interest in forests and climate change. With the failure of international climate negotiations yet to forge an agreement, there are still no formal compliance markets and only a few operating voluntary markets, while carbon prices have dropped precipitously. This means that market-focused carbon schemes are being constructed on a very fragile base, propped up by hype, hope and future expectation. As already discussed, many project developers cannot rely solely on selling carbon credits to meet their costs under current price projections, and business models have had to shift. Increasingly, public support is being required to bolster carbon projects, through government-led REDD+ programmes, supported by the international community and aid/climate finance (Fong Cisneros, 2012), presented now as 'performance-based' aid involving 'public-private partnerships' (Angelsen, 2013). Therefore, like many markets, they are not 'free' as in the neoliberal imagination, but linked to state interests and international public financing.

Of course carbon markets are not the first attempt to commoditize rural forest resources in Africa. These new markets build on previous layers of marketization, pushed at various points by different interests. Thus, for example, in Ghana, timber concessions in the project area had allowed timber trees to be sold off to contractors under the Timber Resource Management Act. The new carbon project was seen very much in this light; indeed, it was initiated by a timber contractor under a similar model. In Sierra Leone, the area now targeted for carbon had previously been a timber reserve. Equally, in Tanzania and Zimbabwe, previous interests in forest and wildlife conservation have coloured the way new markets have developed. In Zimbabwe, the project developers are associated with safari companies and see the fortunes of the carbon project building on, and linked to, protecting wildlife for lucrative game hunting. Their model for community profit sharing is derived from an earlier experience with benefit sharing through wildlife utilization, the famed CAMPFIRE initiative (Dzingirai and Mangwanya, this book). Histories of markets and experiences of commoditization therefore shape how new markets are formed.

As already discussed, all case study projects are planning on selling carbon credits into internationally approved carbon markets, whether through the CDM or the VCS. However, most have found it difficult to get approval. This is a complex, elaborate and expensive process, and only four case examples had gained some form of accreditation by mid-2014 (Table 1.1). The start-up costs of establishing a project have proved especially challenging for project developers, given that carbon revenues are likely to flow perhaps only two or three years after the project is established. This is perhaps the most costly period too, with requirements for surveying the area, evaluating carbon stocks and flows and so on, as well as brokering deals with authorities and local communities. Across our cases, project developers complained that they were finding the going tough, and that external finance was essential. A variety of sources have been deployed, including a mix of public, aid funds, personal and venture capital finance and business sponsorship as part of corporate social responsibility and environmental programmes. This is high-risk financing, and project developers noted that standard forms of business credit and support were not available. However, the 'green' label certainly helps the business proposition, and each of the case study project websites and publicity literature is full of statements about tackling climate change and assuring environmental sustainability.

Most project developers are passionate about the potentials of their projects to protect the environment and the planet. They see themselves at the forefront of innovative environmental management and climate mitigation responses, although often on the basis of a rather simplistic narrative of environmental conservation. Some see themselves as ecological and social missionaries, helping to save poor farmers and their environments. Thus again, the projects and the markets to which they are linked are not operating in the abstract; they are deeply embedded in particular discourses about environment and development, as well as providing profits and business opportunities. While there are inevitably multiple contradictions, and a certain amount of 'greenwash', especially in media campaigns and publicity materials, this socio-political context of market creation is important to appreciate.

The role of the state

Despite the marketized character of forest carbon projects, the state is far from absent. Indeed, carbon makes a difference here too, driving particular interests, opportunities and kinds of state involvement not always seen in other environment–development interventions. Capturing value from carbon depends on a peculiarly large and complex array of social and political relations, in which elements of the state play critical roles.

In general, states are required to uphold the property rights required by markets, as well as to sanction investments and regulate the conduct of operators. A functioning, Weberian bureaucratic state is often assumed in the design of projects, and, just as with 'ideal-type' markets, the assumptions are often found wanting. This is especially true in Africa where post-colonial states have been characterized as clientelistic, predatory, neo-patrimonial and corrupt, representing a 'politics of the belly' (Bayart, 1993; Chabal and Daloz, 1999). Equally, states may proclaim a 'developmental' role – including around carbon investments and climate change mitigation. Yet even such labels are too simplistic. A more nuanced analysis sees states not as singular, unitary authorities, but made of multiple individuals, groups and interests (Boone, 2003; Das and Poole, 2004), each negotiating with each other through complex and shifting alliances and networks (Hansen and Stepputat, 2001). These intersect with the interests of foreign capital and investment, as well as local leaderships, often of multiple, overlapping forms.

Amidst such complex state-business-investment interactions, the ideal-type neoliberal model fostered by carbon projects quickly unravels. But at the same time things can get done, and 'working with the grain', accepting that state politics and bureaucracy do not reflect a liberal ideal makes much sense (Booth, 2011; Kelsall, 2013). Elites, inside and outside the state, become crucial as brokers and negotiators, able to cut deals between state officials and investors, subvert regulations and establish authority – perhaps not totally within the rules but sufficient for it to work. In practice, most carbon projects – indeed most development projects – must operate under such conditions. That these do not meet the full requirements of the assumptions and plans means that the focus must shift to learning by doing. The trouble with carbon projects is that there is limited room for manoeuvre, as the audit and accounting requirements imposed are so strict, so there have to exist parallel worlds of realities on the ground and the idealized plans and proposals. As the case studies show, this can result in severe tensions.

Some particular state interests are at play in forest carbon projects. Central states see new commodities as a source of rent, through licensing, tax and so on. In many African economies, the recent commodity boom has spurred growth on impressive scales. While carbon cannot match gold, diamonds and other minerals, it is seen in the same light, especially in settings where extensive forest resources are deemed 'underutilized'. Bringing the forest – which of course is not underutilized at all in most instances, as forest livelihoods of diverse sorts depend on it (see below) – into the realm of commodity trading means that, like minerals, it can become an important source of state revenue. Carbon funds also potentially create new money in addition to aid, and although, as we have seen, most projects depend on classic development finance at least in their start-up phases, carbon funds offer the prospect of long-term, sustainable finance not subject to the fickle whims of aid flows. Locked in for the long-term, there are real incentives to sustaining the asset, while new carbon finance allows this to happen. Thus, protecting forests and wildlife areas, often neglected in government financing due to swingeing cuts imposed by structural adjustment conditionalities, can now happen, with forest guards employed, fences constructed and management regimes imposed. In Tanzania and Sierra Leone this configuration of state interests and external finance is now reshaping landscapes, recreating the 'fortress conservation' of past times (Kijazi; Winnebah and Leach, all this book), while in Uganda (Nel, this book) the recruitment of private companies to implement carbon projects within forest reserves assists the state with its revenue streams, policing and community relations.

As different departments try to cash in on carbon funds and struggle over technical domains, so carbon projects expose varied interests within the state. Most REDD initiatives are overseen by an environment ministry or a forest department. Suddenly such parts of government are apparently flush with money and the status that comes with external interest and technical authority, with new vehicles, refurbished offices and *per diems* being paid out at endless meetings. This can bring resentment and competition, as in the cases of Zambia and Zimbabwe, where ministries such as agriculture had dominion over the rural areas, and carbon is now seen as competing as a commodity with maize, tobacco or cotton. With all ministries trying to seek investment, the balancing of a forest protection strategy under a REDD programme and an agricultural expansion and investment programme are not easy to square. Such conflicts are in turn played out in patterns of landscape change and control.

At the local level, local governments often become intensely involved – especially in decentralized jurisdictions where they have notional, if contested, control over land and resources. This proved critical in a number of the project cases. In the Mt. Kilimanjaro case, there is a conflict between the Regional Commissioner/Regional Administration and local government. By declaring a 'climate crisis', the former has now taken over oversight of all forest activities, including forest carbon projects, although, by law, forestry activities should be undertaken by local (district and village) governments. This disenfranchisement of local governments and communities is, therefore, at the centre of current forest conflicts in the area (Kijazi, this book).

Local governments may also be cash starved, if funds do not flow from the centre. Thus, for a new carbon project to offer revenue sharing with the local government authority around a commodity that they notionally control ensures that interests are aligned. In the Zimbabwe case (Dzingirai and Mangwanya, this book), for example, the Rural District Council, as the notional land holder of the communal areas, is a partner in the project. The argument is that this allows for local democratic control through councillors, who are accountable to the people, and so benefit sharing is assured. However, this governance arrangement does not please everyone. Elite in-migrants, often with close connections to political party factions, oppose the carbon project, arguing that tobacco growing, which requires forest clearance, is the future. Yet the Rural District Council and carbon project partners have the backing of the chief, who sees carbon as a source of benefit to him personally and the communities loyal to him in a way that tobacco growing by migrants is not. These carbon conflicts, in Zimbabwe as elsewhere, therefore involve multiple competing interests, with the state interceding in different ways and in different forms.

These wider interests are also driven by the potential individual gains to be made from carbon projects. Political elites may form alliances with a project, or against it, depending on how they see their interests being played out. For example, the project developer in Ghana alleged that the ruling party attempted to sabotage the project through denial of financing and later an arrest by the Bureau of National Investigation, because of perceived links with the political opposition (Hashmiu, this book). In Kilimanjaro, Tanzania, the call to preserve carbon is supported by an alliance of the state and a large UNDP/GEF project (Kijazi, this book). This has effectively reinforced existing criminalization of small-scale fuelwood collection, timber harvesting and charcoal making, and promoted the provision of alternative fuels. The latter are provided by richer business entrepreneurs from outside the area, and local villagers lose out. Meanwhile, elites with close connections with state officials are able to subvert logging bans and continue their extractive businesses unhindered. Formal and informal rent seeking among different state agencies attempting to retain climate aid money has led to power struggles in the government bureaucracy. As these have played out, carbon funds and forest governance powers have become concentrated with the Kilimanjaro park authority and the regional administration, opening up the opportunity for corruption and rent capture. This has happened at the expense of democratic governance of forests, as elected district and village councils have lost their powers.

In all cases, although taking different forms, we see such intertwining of party politics, bureaucratic competition, business interests, land and ethnicity, revealing carbon projects as new sites for political struggle. These socio-political conflicts can be seen as an inevitable corollary of carbon projects, given the plurality of diverse interests and social and political relations amongst different local, national and global actors required to realize value from carbon.

This wide array of relations means that carbon projects, often in remote rural locations where state agents are thin on the ground, can take on state-like functions and characteristics. This may involve providing social services (schools, health posts), security (forest guards, poaching patrols) and quasi-democratic functions (consultative committees, community groups), as shown across all the case studies. Projects – with varied mixes of private, NGO and public sector staff and characteristics – may also undertake environmental planning, review, assessment and governance functions that the state might have been expected to assume. Carbon projects can be welcomed as preferable to previous arrangements, as in the case of the Kasigau project in Kenya (Atela, chapter 6, this book). Here people had been excluded from local wildlife resources thanks to a centralized regime and deprived of state services, so many welcomed the project and its associated community benefits. By contrast in Uganda (Nel, this book), the establishment of projects resulted in exclusions and evictions, although in some instances, in these 'zones of awkward engagement' (cf. Tsing, 2005), renegotiations with local company and state officials took place.

Carbon projects – like other forest projects – change relationships of power, representation and accountability, including between communities and the state

(Agrawal *et al.*, 2012; Marino and Ribot, 2012; Ribot and Larson, 2012). They reconfigure local politics and interests, and with this who has control over resources. In some of our cases this results in an essentially private actor, often a private company, gaining control over land and trees that were once in the hands of communities and traditional chiefs – as the Ghana example shows. In other cases, carbon projects act to recentralize power, extending the reach of the central state, over and above local decentralized authorities. We see this in the cases from Tanzania and Uganda, and also Sierra Leone, where the WAPFoR project is reasserting and enabling government control over the forest reserve and, on the grounds that the project is on government land, denying any local rights to participate in shaping project activities.

Landscapes are recast and shaped through such political processes. Yet quite how this plays out is highly dependent on the context, including histories of state formation, patterns of decentralization and forms of elite control. A particularly important factor is the nature of the land – and tree – tenure system in place, a theme to which we now turn.

Tenure: Negotiating rights and access

Carbon markets rely on the trading of property: carbon as a commodity must be owned. Yet the peculiar characteristics of carbon – constructed as a commodity through discursive processes, and subject to diverse imaginaries – complicate its tenure. How, for instance, might one seek to define ownership of 'the mists one sees above forests in the morning'? How are people to define rights and stake claims over 'carbon' when it is so hard to understand, and its value embedded in such diverse and distant relationships? These characteristics configure and add significant ambiguities to the process of negotiating rights and access over forest carbon, adding to those that already pervade tenure systems. For the individual property rights on which many market regimes rely are far from the reality in rural Africa. Instead, tenure systems involve complex, socially and historically embedded mixes of state, communal and individual ownership. Rights over land and trees are layered and overlapping, often contested and usually ambiguous, existing in a pluri-legal setting (Berry, 1989, 2002; Peters, 2004, 2009; Lund, 2008; Sikor and Lund, 2010; Peluso and Lund, 2011).

So how does carbon become property, and so enter markets in the cases we have examined? There are different routes, depending on the type of project (see Table 1.1). In some cases, carbon is 'farmed', as part of agroforestry and tree planting efforts. Thus carbon rights become aligned with individual rights over trees. This may seem simple, given that most such trees are planted on people's individual plots or homestead gardens. Yet tree tenure, even in such settings, is complex, with some trees individually owned, while others may have access and use rights: for fruits, non-timber forest products, or social activities (Fortman, 1985; Unruh, 2008).

In other projects, the carbon is enclosed as part of a 'ranching' or 'fortress' conservation arrangement. A marked boundary or large fence is put up around the area, and is guarded. These sites – as in Kasigau in Kenya, LZRP in Zambia, WAPFoR in Sierra Leone or the Uganda examples – are former large-scale farms, ranches owned individually under freehold title arrangements, or conservation or forest areas, demarcated as state land. Here carbon rights are linked to land rights. But again these may be less clear than first imagined, despite obvious demarcation and cadastral authority. Other competing rights may exist that cut across the fences, supported by claims based on earlier periods, and confirmed by customary law and spiritual authority. Grave sites, past village settlements, particular trees, wetlands or caves, for example, may be used as evidence that others have rights over the areas, beyond those signalled in state land registries or proclamations.

These competing claims are often accepted by the formal land owner as part of a local social deal that allows, for example, rainmaking ceremonies at sacred sites, visits to burial grounds at particular times of year, and the use of the area for harvesting of forest products or other livelihood activities. So, for example, in Kasigau in Kenya, people are permitted to attend sacred shrines and grave sites, even though these are on hills within the project boundary. In Mt. Kilimanjaro, people have since colonial times had access to the forest reserve for livelihood and cultural uses – collecting fuelwood and livestock fodder, accessing traditional ceremony sites, collecting medicinal herbs or repairing indigenous irrigation furrows that originate in the forests. They view such access as their customary right.

Such social access arrangements redefine the sharp lines of forest boundaries and cadastral surveys into something much more fluid. However, such fuzzy boundaries recognizing overlapping claims are poorly aligned with the imperatives of carbon projects, which must protect carbon value as exchangeable private property and assure its presence for decades. Land holders and carbon beneficiaries are therefore forced to rethink the security of their resource, making it only rational to reassert the boundaries and create a more fortress-like arrangement that excludes others' access or regulates this with much more rigour. We see this in the Kilimanjaro case, where longstanding access has recently been curtailed by new carbon initiatives: all men have been banned from going into the forests, while women have been given very restricted access. Indeed for some, the requirements of carbon projects, and the funds that follow, allow for the imposition of long-desired controls. While a more community-based approach may have been the default in the past, partly the consequence of lack of funds to do anything else and partly because joint management approaches were heavily backed by donors, officials have often hankered after a more traditional exclusionary, often militarized, approach to conservation. In the Sierra Leone case, this is exactly what has happened, with the WAPFoR project enabling a long-desired major expansion in the numbers of forest guards, now armed and given police support.

Another type of project aims to appropriate carbon value from community held resources. This presents some of the most challenging tenure issues. Across our case studies, 'communal' land is held in different ways – as part of chiefly 'stool' land with minimal central state interference (in Ghana), as a hybrid arrangement between traditional authorities and the local state (as in Kenya, Tanzania, Zambia and Zimbabwe) and as a patchwork mix between private, community and state land, with unclear boundaries. The Kenyan KACP case (Atela, Chapter 4, this book) represents a situation whereby the carbon accounting procedure is based on individual land holdings, linked to 'carbon rights' and associated payments. However, more than half of the land in the project area is held customarily and legitimized by traditional passage of use rights from one generation to the other. Customary land rights are held by individual families, but land is often used communally. Given that residue incorporation and vegetation retention in these farms are some of the key carbon-generating activities, should farmers allow communal grazing of land during the dry season or instead conserve residues for sequestration and individual benefit? Such conflicting land and resource tenure arrangements may create significant social conflicts as the commoditization of carbon creates incentives to privatize and individualize resources.

Deals made with 'communities' thus present real challenges, as it is often not clear who the community is, what authority the notional leadership has over land and resources, or how such a community may change over time. Contemporary arrangements of course reflect past histories, and especially particular patterns of colonial rule. Thus in countries where indirect rule dominated, such as Ghana, chiefly authority is significant; deals that stretch across different stool areas are problematic, as are commitments by the central state (Kasanga and Kotey, 2001). In Sierra Leone, parallel situations prevail across most of the country, but the Western Area where the case study project is situated is an exception: this was once a British Crown Colony, and all land is still formally owned by central government. In former settler economies, such as Kenya and Zimbabwe, there was once a clear division between white-owned freehold land and communal land (the reserves, or tribal trust lands), although the state had jurisdiction over these areas, allowing chiefs at different times some level of control over land allocation and adjudication (Berry, 2002; Peters, 2004). Tanzania and Zambia represent situations where land is held by the state (sometimes through local government under decentralized arrangements), but traditional authorities have been granted substantial control on customary or 'tribal' lands. Depending on wider politics and the power and imperatives of the state, these arrangements vary over time. Land rights and markets involve many 'vernacular' interpretations (Colin and Woodhouse, 2010). 'Traditions' – and with these customs, practices, rules, regulations, and community forms of authority – were, and still are, often 'invented' to suit particular circumstances (Hobsbawm and Ranger, 1983; Chanock, 1991). When a new carbon project arrives it must slot into this layered history and experience, negotiating access and authority accordingly.

Forest landscapes are in part accretions of past tenure relations. Past authority can be reinvoked if new forms of control are imposed that affect particular interests, or new forms of value are realized. We see this in many of the cases, in the context of disjunctures between project assumptions and imperatives around state or individual control over carbon, and complicated tenurial realities. Suddenly a long-abandoned grave site becomes a central bone of contention, or previously hidden forest uses become criminalized and resentments rise, or past disputes between different chieftaincies or headmen become regalvanized, as it suddenly becomes important to assert authority over once peripheral but now newly valuable areas. A landscape always has a long, layered history replete with different memories, meanings and claims. Sometimes new disputes revive even pre-colonial histories, with conflicting assertions of power and control over an area based on different groups' often highly stylized unwritten versions of the distant past.

Given these multiple jurisdictions and types of authority, and competing, layered histories of claims in African rural settings, it is not surprising that conflicts over land and resources are perennial, continuously negotiated and renegotiated, with outcomes depending on the power relations between the different actors (cf. Berry, 1989, 1993; Peters, 2004, 2009; Sikor and Lund, 2010; Peluso and Lund, 2011; Lund and Boone, 2013; Boone, 2013). Carbon projects not only find it difficult to navigate these complexities and conflicts, but often fuel them, adding further dimensions. Thus, for example, in the Ghana case, the project introduced new actors by leasing land and carbon rights to 'absentee owners' – including urban-based professionals, church groups and others with no historical relationship to the area (Hashmiu, this book). In Kasigau, Kenya (Atela, chapter 6, this book), the project in effect dismantled the state-based institutions for land control and imposed new ones, adding a new layer of jurisdiction and multiplying ambiguities. In Zambia, the strengthening of private property rights for carbon acted to weaken customary controls over land, replacing local institutions. This opened the door to other forms of 'land grabbing', making it easier for private investors to come in and make deals with the local state (cf. Fairhead *et al.*, 2012). The converse can also happen. In Zimbabwe, the alliance of the carbon project with local traditional leadership and decentralized state authorities has had the effect of preventing land grabbing by outsiders, and the project was seen by local inhabitants as a way of halting the influx of migrants into the area, thus protecting 'indigenous' rights (Dzingirai and Mangwanya, this book).

In these ways, questions of property rights and tenure are at the heart of carbon conflicts. Carbon interventions (re)shape resource access and control, and interplay with the politics of land and trees in ways that create and legitimate new forms of social order, and contestations over authority (cf. Lund and Boone, 2013). Conflicts unfold as carbon projects attempt to impose the neat market-based property arrangements that they require onto complex, ambiguous, rural tenure discourses and practices, and the ambiguities of carbon itself as a commodity. Forest landscapes are re-shaped in the process. But this happens in very varied ways, depending on historically derived configurations of power and authority in particular areas, and is open to ongoing negotiation. The new forms of control that are derived from forest carbon projects therefore are not all-powerful: the unruliness of people and politics, as well as of nature, can undermine neat plans. Nor are the winners and losers predictable; this reflects multiple factors and the particular diversity of other interests at play. Similar variation arises when, as in the next section, we reflect on the differential impact of carbon forestry projects on livelihoods.

Diverse livelihoods and styles of resistance

In the sometimes rather simplistic rhetoric around the assumed 'pro-poor' impacts of carbon projects, there are narratives about 'benefit sharing', 'community-based' impacts and 'empowerment'. These are also central emphases in the safeguards and standards adopted, at least in principle, by many carbon project frameworks. Such narratives have become standard in development more generally, but as the widespread critique of community-based resource management has shown (Nelson, 2010; Dressler *et al.*, 2010), simplistic assumptions have to be rejected in favour of a much more differentiated view. Communities are not uniform, but intersected by diverse axes of difference – from wealth to gender, age and ethnicity. Diverse institutions are associated with powers of inclusion and exclusion, shaping resource access and control (Ribot and Peluso, 2003; Hall *et al.*, 2011). Analysis of livelihood impacts must attend centrally to such institutional and political processes (Scoones, 2009).

In these respects, the analysis of livelihood impacts – and sometimes resistance to negative impacts – in our cases reveals many continuities with past rural development interventions. Yet here too, carbon makes important differences. The business-oriented character of many projects, and the idea that profit is being made by distant outsiders, brings a greater sharpness to local debate about who is gaining and who is losing. Carbon projects typically involve new development actors unfamiliar with past lessons. And the ambiguous, hard-to-fathom processes that give value to carbon both fuel local anxiety, and make resistance hard to focus.

In this context of continuity and change, the cases find that winners and losers are not necessarily as expected. In none of the cases does 'the community' line up against, or indeed with, 'the project'. Rather, there is a diversity of livelihood interests and ways of relating to forests and ecosystems embedded in people's settlement histories and social positions. Sometimes unexpected alliances form. Thus, in Zimbabwe, support for the project comes from indigenous food crop farmers and their chiefs, as a way of keeping at bay rapacious migrant tobacco and cotton farmers (Dzingirai and Mangwanya, this book).

Project designs often overlook such differences, either casting communities as unitary or adopting rather simplistic views of who are the 'culprits' and who should be the 'beneficiaries', in ways that leave certain groups excluded. Thus, for example in Ghana, migrant sharecroppers may not own land but are important land users; however shifting land to a carbon reserve unwittingly displaced them (Hashmiu, this book). In Zambia, migrant charcoalers are perhaps the major threat to carbon resources. Yet such itinerant livelihoods are difficult to understand and control by a project focused on 'community' structures (Mickels-Kokwe and Kokwe, this book). In eastern Kenya, transhumant pastoral livelihoods were disrupted by a project focus on resident agriculturalists, despite the fact that both had claims on the area (Atela, Chapter 6, this book).

In projects that involve the allocation of farmland for tree growing, resource access and control arrangements mean that only some people can participate. This is usually larger landowners who can afford to allocate portions of their farms to trees, while retaining areas for food crops. Very often it is male land 'owners' who participate, and women, who manage smaller garden areas, do not directly benefit, while those without land, including youth, hoping to inherit portions of their fathers' plots, are disenfranchised. This was the case in Ghana (Hashmiu, this book), where older indigenous men could afford to allocate land to carbon trees, whereas women, youth and migrants needed the land to grow food crops. Thus, carbon projects can have gendered and age-specific consequences, with the value being appropriated by some but not others.

The cases illustrate many other forest-linked livelihood activities, important to different people – from beekeeping, foraging and small game hunting to dry season or drought relief cattle grazing, and selective timber harvesting for house construction or boat building. As recent studies have shown, the value of forests beyond carbon can be considerable (Sunderland *et al.*, 2014; Jagger *et al.*, 2014). Small-scale extraction may not jeopardize the carbon stock significantly, but in forest-savanna and dry forest zones the threat of fire often means that project developers act to exclude such people. New security arrangements – fences, guards and anti-poaching patrols – act to criminalize such livelihoods, creating resentment and conflict. This may have a contradictory effect in practice. For instance, if fire management is not taken seriously, the build-up of grass in the absence of grazing or controlled burning may create a major risk to carbon stocks – a live issue in the Ghana case study (Hashmiu, this book). Equally, by making foraging, hunting and grazing illegal, those who continue to practise such livelihoods do it under cover. In order to flush out game, create patches of grazing and clear areas rapidly, fires may be set, causing more damage. Thus, new regulations, creating new forms of exclusion and changes in livelihood use, may change the ecology, and so restructure the landscape.

Accepting that carbon forestry projects must protect their newly valued resources, offering alternative livelihood options that do not affect the carbon stock is invariably part of project designs.[2] Indeed, showing that such alternatives exist is crucial to the argument that carbon is being stored at higher levels than would have happened without the project. Thus, across our cases there is an array of alternative livelihood interventions, most part of standard development repertoires. Some involve alternatives to using a protected forest area, such as beekeeping and hive construction or mushroom growing in village areas; some involve reducing forest destruction, including improved-efficiency technology for stoves, brick making, bread ovens, fish smoking, tobacco curing and charcoal manufacturing; some involve changes in agricultural practice, including 'conservation agriculture' or agroforestry, that allow for increased carbon sequestration or that intensify agricultural production (such as improved irrigation technologies) or that increase the value of farm production (such as negotiating better prices); others provide new livelihoods that are not dependent on forests, including the financing of motorbike taxis, and the inevitable chicken, garden and craft projects for women; and finally, others are focused on gaining community acceptance, through investments in building schools, clinics and grinding mills, as part of a corporate social responsibility/public relations drive.

As the case studies show, such livelihood interventions face persistent problems. These are not exclusive to carbon projects, reflecting the long, hard experience of rural development more generally. However, very often new carbon projects are run by those with limited rural development experience, so old mistakes are repeated. Thus activities sometimes become captured by particular elites, entering local political struggles. Some fail to produce a viable source of income and therefore do not replace carbon-consuming alternatives. Communities, as noted, are far from uniform, and so activities geared to assumed 'community needs' fail to articulate with people's livelihood priorities. Activities sometimes assume particular gendered and age requirements based more on stereotypes than real aspirations, and so are rejected. And they may require substantial labour that in reality is not available, particularly for smaller households, those with young children, the aged, sick or infirm. Thus livelihood activities can miss their target of being 'pro-poor', by failing to understand the different constraints of the differentiated poor within an area. The case studies illustrate many such instances. Most fundamentally, though, they show that due to false narratives and misperceptions about forest use and change, and their interrelationships with 'carbon', livelihood interventions are often based on illusory premises. The design assumptions of increasing carbon sequestration by often significant percentages through such efforts are, as we see in the cases, therefore way off the mark.

The disconnects between project promises and livelihood impacts, as well as local experiences of resource appropriation, can lead to deep local resentment. The resulting carbon conflicts take on different forms, from outright challenge to a project to more passive, hidden forms of resistance (Scott, 1990). Of the cases in this book, the Tanzania case showed the most overt forms of resistance, as villagers mobilized to demonstrate against coercion associated with fortress forest protection, particularly the use of violence by park rangers. This went as far as digging trenches in the road to stop tourist vehicles entering the area, along with suspected arson and sabotage within the now enclosed forests (Kijazi, this book). In Sierra Leone, equally, aggression towards forest guards has been reported, as people become frustrated by the imposition of the new boundaries to the forest reserve (Winnebah and Leach, this book). Other cases show a pattern of outward public acceptance, but continued foot-dragging that makes a project difficult to implement. Local negotiation and accommodation may take place, where plans are amended, rules relaxed and informal practices accepted; yet these may only emerge where flexibility and discretion is permitted for project workers (Nel, this book).

Public media have been used in both promotion and resistance to projects. Project developers are quick to highlight that their efforts are pro-poor and green in their public proclamations, even styling themselves as 'missionary' ecologists and developers (Dzingirai and Mangwanya, this book). Many use high-profile community investments – such as schools or clinics, opened with fanfare – as public demonstrations of project value. In their public relations efforts, projects also seek to distance current activities from past interventions that have foundered and gained a bad reputation. Thus, in Zimbabwe, the Kariba REDD project is at pains to point out that, despite obvious similarities, it is different to the wildlife utilization CAMPFIRE schemes of the past that became riven with conflicts and accusations of improper fund appropriation and lack of benefit sharing (Dzingirai and Mangwanya, this book). But resistance may also deploy similar tactics: highlighting similarities with past failures, and spreading rumours about projects' 'real' intentions. In Zimbabwe, local people used dramatic terms, arguing 'they are here to kill us', and playing into the racially tinged politics of such efforts. In Sierra Leone (Winnebah and Leach, this book), journalists and NGOs have used newspapers and radio to associate the WAPFoR project with 'carbon cowboys' – companies allied with allegedly corrupt state officials who have developed schemes elsewhere in the country geared to profit at the expense of communities. Carbon projects in Africa are often interpreted by publics amidst wider media discussion of corruption, scams, and, in the West African context, accusations of projects being '419s' (a colloquialism for a fraudulent arrangement invoking this section of the Nigerian criminal code).

These are of course similar to responses to coercive environment and development projects of the past. But what difference does carbon make? One differentiating factor is the casting of carbon projects in such global, generic terms, linked to narratives about global climate change and hard-to-grasp valuation of 'carbon', that they are more difficult to apprehend and respond to at the local level. Another is that benefit flows are over such long time frames compared with the day-to-day and seasonal nature of livelihood decisions. Thus, the high-sounding objectives of carbon projects are often meaningless to local people. This creates situations in which the project and local objectives often fail to connect, increasing the potential for conflict. Yet it can be difficult for local people to know what, exactly, to resist; certainly not the need to do something about climate change – with which most people, if asked, concur. Instead, conflicts themselves tend to focus on day-to-day project practices and livelihood impacts.

Thus, overall, carbon projects have diverse impacts on livelihoods, resulting in a variety of conflicts and styles of resistance. Carbon conflicts take on many forms, depending on the historical and political context, experiences of past interventions and the form of appropriation that the carbon intervention takes, and so the changes in authority, tenure and access that result. Carbon projects thus reconfigure social, political, ecological and even cultural and symbolic dimensions of landscapes in complex ways. Understanding carbon conflicts therefore requires an integrative approach – as taken by each of the chapters in this book.

Before moving to the cases, however, we want to conclude this opening chapter with some reflections, emerging from across the case studies, on future policy scenarios.

Conclusions

There are a number of possible responses to the findings of the cases in this book, and the broad implications laid out in this chapter. In this concluding section, we outline three.

The first argues that carbon forestry projects will inevitably fail due to their inbuilt contradictions. Attempts at appropriating value from carbon that is already owned and used by others is bound to be resisted. Carbon projects are in other words 'green grabs' (Fairhead *et al.*, 2012), whereby carbon is appropriated notionally for environmental ends, but in practice to meet commercial imperatives.[3] Such a 'grab' alienates land and resources, reduces access and results in inequalities within and between groups. The result is inevitably struggle over authority and a contest for benefits between different players. The relationships between land, territory, identity and citizenship are challenged, with major political and social implications. The consequence of any such intervention will, almost inevitably, be conflict, no matter what 'livelihood diversification' activities and 'benefit sharing' options are offered. With the failure of the pure market-based approach to carbon project development due to declining prices and high start-up costs, arguments to treat carbon forestry as an aid-funded 'public good' initiative run by private sector investors appear fraught with contradictions. While few deny the imperative of addressing climate change, the offset market approach through carbon forestry projects in remote rural locations in Africa is so prone to failure through leakage, lack of permanence, project failure and resistance by unruly ecologies and people, that justifications that the global public good of carbon emission reductions should override local costs to livelihoods look extremely shaky. Instead, mitigation at the point of emission may make much more sense if climate challenges are to be addressed. The environmental consequences of excess consumption and un-trammelled growth cannot be addressed in Africa, where other developmental challenges exist. Thus, the argument goes, the apparently neat 'neoliberalization of nature' discourse, based on processes of financialization and marketization, at the heart of carbon forestry projects, is deeply flawed, results in inequitable outcomes and may in fact not deal with climate change effectively anyway. The conclusion under this first response would be to abandon carbon projects and revert, as many indeed have done, to more traditional livelihood and rural development projects that support sound forest management, but with local interests at the heart of design and implementation – and leaving climate change mitigation to be addressed elsewhere.

A second response accepts that high-level international policy concern around climate is mobilizing considerable resources for forests and rural development in unprecedented ways, and that this needs to be capitalized upon. The climate–forest intersection is therefore a major opportunity to develop innovative solutions that benefit local people, improve the sustainable management of forest resources and boost climate mitigation efforts. The role of offset schemes and private sector actors also brings in new mechanisms, expertise and funding into a sector that has been in the doldrums. Rather than taking a pessimistic view, 'win-win' options should actively be sought out. This means bringing lessons for project design developed in previous eras – around community-based forestry, joint forest management, on-farm agroforestry and so on – to bear, but making them more 'climate smart'. As with previous experiences, issues of forest governance – and tenure in particular – are essential components, so a more sophisticated, socially informed approach is required. This means more community participation in planning and design and an approach to decentralizing and managing benefit-sharing so that disenfranchisement, alienation and exclusion do not happen. This response represents a pragmatic approach that accepts that climate-driven projects are a feature of the current environment-development context, but aims to adapt projects and policies to bring governance and participation issues more centre stage, so that these new interventions result in wider benefits for rural livelihoods.

A third response takes a different stance, and derives from a rather different starting point. It accepts the critique of the first response, and so argues for a need to address head-on the politics of marketization and financialization, avoiding a naïve acceptance of market-based offset solutions. But it also takes a pragmatic stance in suggesting that there are things that can be done, as long as we accept that all players are political, and that conflicts over carbon, just as any other resource, are inevitable. Climate finance for mitigation – through a range of market and non-market mechanisms – provides, such a response argues, a useful route to investing in new patterns of resource use, if a rather more acute political analysis of winners and losers, inclusions and exclusions takes place. This third response then starts from the concept of landscape with all its social, political, cultural and ecological dimensions, and asks 'what should this landscape look like in a world where carbon matters?' And it starts from a recognition of 'carbon' not as a commodity like any other, but open to radically different imaginaries. This is different to starting from a landscape as a rational planning tool, carbon as a commodity in a notional market or from a project developed in abstraction. Lived-in landscapes, as discussed earlier, have histories, layered experiences and interventions, and so must be addressed holistically, and through a socio-political lens. At stake may be radically different worldviews – encompassing different meanings of carbon, and priorities around global versus local, or profit versus livelihood, concerns. An inclusive and deliberative approach would have to be at the centre of such an approach; one that is cognisant of such differing views, interests and politics. This would involve discussing diverse futures, from the standpoints of different people and things; debating the views of women and men, elders and youth, richer and poorer, the state and local groups, as well as finding ways of bringing nature and the broader planetary environment into the conversation.

By thinking about future pathways for a particular landscape, trade-offs, obstacles and existing and potential conflicts would be incorporated. The aim would be to construct multiple 'imagined' landscapes across scales and over different timeframes, and so generate different, perhaps conflicting but sometimes compatible, pathways to sustainability. Asking the question 'what do future carbon landscapes look like?' opens up the debate about imaginaries, livelihoods, resource use and political dynamics, linking local contexts with global issues. In this sense, carbon, almost because of its abstraction, can act as a valuable discursive commodity, a boundary object, around which debates around local sustainability within planetary boundaries (cf. Rockström *et al.*, 2009) can unfold, but always highlighting that the negotiation of such pathways will be intensely political (Leach *et al.*, 2010). Carbon conflicts, therefore, should not be a surprise that challenges neatly laid plans, but should be embraced in an agonistic politics (Mouffe, 2005) of dispute and deliberation around resource use and rural livelihoods (Holmes and Scoones, 2000). Such an approach would clearly be a radical departure from the status quo, and would mean a fundamental rethinking of the carbon forestry approach. It also implies a rather different perspective on 'landscape' than the managerialist notions currently being peddled. However, it does pick up on and draw experience from a long tradition of critique of market-based and top-down planning solutions, as well as novel, participatory practice in resource management and rural development (Scoones and Thompson, 1994).

The aim of this book is not to provide neat, definitive policy recommendations, but to unpack fast-unfolding experience in a number of cases and draw some rather wider, analytical lessons. This chapter has attempted to do this, drawing on all the case studies that follow. The cases in different ways suggest elements of all three of the responses highlighted above, depending on the context. Certainly overall, the critique of the simplistic offset market approach comes across loud and clear throughout the cases, and so there is wide support for the first response outlined above. But authors do not go as far as rejecting carbon forestry interventions out of hand. That said, there is great caution observed for the reformist stance of the second response. The real challenges of taking politics seriously are highlighted again and again, and simplistic approaches to 'participation' and 'governance' fixes to replace a 'technical' or 'market' fix are rejected. It is therefore the third response that resonates most clearly with many of the contributions of this book. The 'politics of carbon landscapes' approach has yet to be fully elaborated, and certainly remains to be tested on the ground, but it offers some prospect, drawing on long experience in a range of fields, for revitalized carbon forestry approaches where carbon conflicts are emphasized in the context of a political ecology approach to livelihoods and landscapes.

Notes

1 See: www.redd-monitor.org/2014/01/09/global-carbon-markets-have-shrunk-in-value-by-60-since-2011/; www.redd-monitor.org/tag/boiler-room/ (accessed 26 May 2014).
2 http://blog.cifor.org/21257/are-alternative-livelihoods-projects-effective#.U4Mk A01OXIV (accessed 26 May 2014).
3 www.redd-monitor.org/2014/03/20/redd-could-lead-to-a-carbon-grab-new-report-from-the-rights-and-resources-initiative/ (accessed 26 May 2014).

2

FOREST CARBON PROJECTS AND POLICIES IN AFRICA

Albert Arhin and Joanes Atela

Introduction

In 1992, the United Nations Framework Convention on Climate Change (UNFCCC) was created to serve as the basis for a global response to meeting the climate change challenge. Since then, the body has spearheaded several agreements including its flagship, the Kyoto Protocol. These are aimed to meet the Convention's core objective of stabilizing greenhouse gas concentrations in the atmosphere at a level that will prevent dangerous human interference with the climate system. In this context, conserving or enhancing forest carbon stocks is considered both as a way to reduce carbon emissions from deforestation and, most importantly, to offset carbon emissions in developing countries.

However, the idea of solving climate challenges through carbon markets remains one of the controversial areas within the UNFCCC policy framework. This is seen in the ongoing global efforts to address climate change through the mechanism for reducing emissions from deforestation and degradation plus the role of conservation, sustainable forest management and enhancement of carbon stocks (REDD+). Despite such heated debate, there are now more than 400 carbon forestry projects ongoing across the world (Gallemore and Munroe, 2013).

What have been the main justifications – in the midst of such controversy – behind the upsurge of forest carbon projects in Africa? What has been the experience of Africa in the participation in the different market schemes for carbon projects? And what can we say about the prospects of carbon projects becoming part of mechanisms to create pathways for achieving goals of poverty reduction, social justice and improvement of livelihood needs of local communities?

This chapter provides an overview of the theory and practice of forestry carbon projects, with particular emphasis on Africa's participation in these schemes. The chapter builds on earlier work by Fong Cisneros (2012) that usefully mapped different forest carbon projects in Africa. It starts with an overview of the institutional and market mechanisms in forest carbon projects, and continues with an update on the state of play in Africa. The chapter concludes with a discussion of the trade-offs involved between market-based carbon projects and wider objectives of sustainable development.

Forest carbon projects: Design features

Forest carbon projects have been gaining traction in recent years within the global environmental governance arena and particularly in Africa. Acronyms fly in a bewildering mix: VCS, CCBA, REDD+, CDM, JI, AFOLU, LULUCF, WB, UN-REDD, FCPF, CERs, VCU, ETS, OTC, A/R, COP, CMP, DNA, DOE,[1] Annex 1 countries, Non-Annex 1 countries and many, many others. Forest carbon projects are initiatives involving payments or funding from conserving, protecting and establishing forestry and agroforesty landscapes that capture and store carbon, which leads to reduction in greenhouse gases (GHG). Forest carbon schemes are part of global efforts to mitigate climate change (Peskett et al., 2011). Generally, such projects must be able to prove that the cut in emissions is real, permanent and verifiable. Plans on how this proof is to be achieved must be provided in the form of a Project Design Document (PDD) and activity reports are subsequently validated by an approved third party or standards.

Forest carbon projects are a sub-set of 'Payments for Ecosystem Services' (PES) projects which gained prominence in the 1990s as one of the policy planks to improve the efficiency of natural resource management (Pagiola et al., 2005). PES generally refers to a voluntary transaction where a well-defined ecosystem service (ES) is 'bought' by an ES buyer from an ES provider (Wunder, 2005). PES schemes, so the theory goes, make it possible for non-market values to be translated into real financial incentives that are expected to result in payment to local actors who are reliant on natural resources to conserve landscapes and secure ongoing provisions of ecosystem services (Engel et al., 2008; To et al., 2012).

These principles underlying PES have influenced the institutions and markets that have emerged around forest carbon projects. A set of technical and institutional requirements exists to create a 'carbon project' (Leach and Scoones, 2013). Although variations exist, depending on particular markets and standards, there are a common set of characteristics that are outlined below.

i   **Defining eligible lands**: In order to fit relevant project classifications, projects have to show that the area complies with an accepted definition of 'forest'. Land eligibility differs among different carbon markets. Under the Clean Development Mechanism (CDM) for instance, if the activity involves reforestation, the land must have been categorized as non-forest as of 31 December 1989 and at the beginning of the project. If the project involves afforestation, then the land must have been categorized as non-forest for the past 50 years. For developing countries, demonstrating eligibility and finding available land of suitable size has often proved difficult (UNEP, 2010). Land eligibility can often be an obstacle if projects cannot provide the appropriate documents, approvals, contracts or whatever is needed to establish proof of title or right of use (Seifert-Granzin, 2011).

ii **Establishing baselines and reference levels**: Projects are required to establish reference levels or baselines which performance will be measured against. The establishment of a baseline should include a change scenario which explains what would happen if no project intervention took place, as well as a reference area. Baselines often assume patterns and trends of deforestation, based on imperfect data sources, especially in Africa.

iii **Proving additionality**: Projects must prove that they will produce carbon benefits which are additional to what might have happened anyway in the business-as-usual scenario (Angelsen, 2008). This entails providing a benchmark for assessing the impact of carbon or REDD+ interventions.

iv **Limiting leakage**: Leakage is measured as increased emissions or reduced sequestration outside the project that is caused by the implementation of the project and which must be deducted from the total amount of certified emission reductions (UNEP, 2010). By addressing leakage, project developers are generally required to ensure that reduced emissions from deforestation and degradation in one geographical area do not lead to negative impacts and higher emissions in another area.

v **Addressing permanence**: Permanence means ensuring that carbon sequestered is not released back into the atmosphere (UNEP, 2010). This has been one of the most controversial issues since the early days of the Kyoto Protocol due to the non-permanent nature of forest carbon itself (Angelsen, 2008). By addressing permanence, project developers are expected to prove that forest area saved today will not be destroyed tomorrow.

vi **Monitoring, reporting and verification (MRV):** MRV is the monitoring, reporting, and verification of GHG-emission reductions/removals from forests using field measurements, remote sensing or modelling. This includes estimation of the extent of significant land use classes and monitoring of land use change within and between various classes. Thus, project developers are required to demonstrate how they would measure reductions in emissions and through which data.

vii **Promoting co-benefits:** Co-benefits are benefits generated by a forest carbon project beyond the emissions reduction benefits, especially those relating to social, economic and biodiversity impacts (Brown et al., 2008). They incorporate aspects of poverty alleviation, livelihood security, strengthening of local community rights and conservation of biodiversity, soil and water into the project design with the aim of enhancing sustainable development in the project area (Greiner and Stanley, 2013). Here, project developers need to demonstrate how they can maximise co-benefits and avoid doing harm to local communities.

As discussed in Chapter 1, the combination of these requirements for forest carbon projects acts to push the design of interventions in particular ways. The complexity of registration and approval requirements is such that projects require substantial inputs from designers, project developers and consultants.

Trading carbon: Market mechanisms

The forest carbon economy is an interconnection of schemes with different institutional arrangements and markets. These take different forms in different parts of the world. The markets trade reduced emissions from forestry schemes in the form of carbon credits. The two main markets in which (forest) carbon credits from Africa and elsewhere are traded are (a) the compliance markets and (b) the voluntary markets.

Compliance markets

The Kyoto Protocol created three flexible financial mechanisms which make up the compliance market. These include (i) the CDM (which makes it possible for Annex 1 or industrialized countries to meet their emission reduction targets through investment in developing countries or non-Annex 1 countries); (ii) Emissions Trading (involving international transfer of emission allocations between industrialized countries); and (iii) Joint Implementation (where any industrialized country can invest in emission reduction projects in any other industrialized country as an alternative to reducing emissions domestically). The CDM is the only flexible mechanism that involves both industrialized and developing countries, and is the focus for this chapter.

The CDM was created to serve two main goals, namely, to assist Parties not included in Annex 1 in achieving sustainable development and to assist Parties included in Annex 1 to meet their emission targets at a lower cost. Although it started operating in 2001, it fully became functional only in 2005 when the Kyoto Protocol entered into force. Under the CDM, emission reduction projects in developing countries can earn carbon credits, termed certified emission reduction credits (CERs) which are measured in tons of CO2-equivalent. Buyers from developed countries acquire Certified Emission Reductions (CERs) for each tonne of GHG that is prevented from entering the atmosphere as a result of a CDM project in a developing country. These saleable credits can then be used by industrialized countries to meet part of their emission reduction targets under the Kyoto Protocol.

Unlike cap-and-trade or allowance trading, in which Parties are granted a quota of emissions that they trade under this cap, the CDM is a project-based approach, with new credits continuously being created as new projects are verified (Boyd *et al.*, 2009). A CDM project is required to provide evidence that emission reductions are additional to any reductions that would have occurred without the project. Thus, projects must, as a necessity, pass through a rigorous public registration and issuance process and also be approved by Designated National Authorities, which are the organizations granted responsibility by a Party to authorize and approve participation in CDM projects.

The host country for the investment must be a developing country that has ratified the Kyoto Protocol. The CDM is subjected to the authority and guidance of the Conference of the Parties to the Kyoto Protocol (CMP) although its operation is managed and supervised by the CDM Executive Board (EB), which is composed of ten representatives of governments from developing and developed countries that are party to the Kyoto Protocol, as well as a number of authorized observers. Once a CDM project has been documented, it must usually be approved by an organization – called a Designated Operational Entity – accredited by the EB. This process is known as validation, the completion of which leads to formal registration of the project and issuance of credits.

In the forestry sector, the Kyoto Protocol limited itself to only afforestation (establishment of forests on lands previously unforested) and reforestation (establishment of forests on lands previously forested, but deforested as of December 1989) and excluded forest conservation and emissions reduction from deforestation and degradation from its first commitment period (2008–2012). Despite the many efforts from a wide range of international, regional and national organizations, the contribution of afforestation and reforestation to the generation of carbon credits under the CDM has been negligible. To date, just less than 5 per cent of CDM projects registered have been from the forestry sector (Boyd et al., 2007; Thomas et al., 2010). Further, the CDM did not consider the crucial role of agriculture as part of forestry carbon schemes through agroforestry.

Voluntary carbon markets

The voluntary market is an alternative to the compliance market for developing countries, including in Africa. In the voluntary carbon market, voluntary emission reductions (also called voluntary emission units) can be acquired by public or private entities interested in voluntarily offsetting their emissions. The term 'voluntary' refers to the fact that players participating in this market are not necessarily constrained or inhibited by emissions targets (which may force them to trade for credits elsewhere); instead, they do so voluntarily to offset their emissions due to factors such as philanthropy, corporate social responsibility, investment in cleaner technologies, preparedness for eventual change in policies and ethically informed management of climate-change impacts (Benessaiah, 2012; Peters-Stanley and Yin, 2013; Portaccio et al., 2013). Buyers include a wide array of actors such as international non-governmental organizations (NGOs) for conservation, governments, indigenous communities, and private actors. Sellers participating in the voluntary market can be categorized into three major types: (a) project developers and proponents that develop GHG emission reductions through offset projects; (b) aggregators or wholesalers that sell offsets in bulk and often have ownership of a portfolio of credits; and (c) retailers who sell small amounts of credits to individuals or organizations, often online, and have ownership of a portfolio of credits (Hamilton *et al.*, 2007, 2008). Sellers or suppliers in the voluntary or offset market thus include (international) conservation organizations hoping to harness the power of carbon finance, retailers selling offsets online, project developers primarily interested in generating VERs and developers of potential CDM projects with credits that, for a variety of reasons, cannot currently be sold into the regulated markets (Hamilton *et al.*, 2008; Portaccio *et al.*, 2013).

Following the need to certify and validate the emission reductions from the voluntary markets, several certification bodies have emerged. The Verified Carbon Standard (VCS), a private standard developed by the International Emission Trading Organization, the Climate Group and the World Economic Forum Global Greenhouse Gas Register, has become the global benchmark standard for project-based voluntary emission reductions that provides a degree of standardization. Other emerging internationally recognized standards that certify voluntary carbon projects include the Climate, Community and Biodiversity Alliance (CCBA), the Plan Vivo, the Gold Standard and the Carbon Fix. Voluntary buyers generally place high preference on projects embodying sustainable development benefits, including improvements in livelihoods, and, as a result, have attracted large numbers of forestry projects. Over the past few years, a growing understanding that forestry and land-use carbon offset projects are inextricably linked with communities and biodiversity has resulted in some partnerships among standards (e.g. VCS+CCB, Gold Standard+FSC+Fairtrade). In 2012 two-thirds of all transacted offsets tied to a standard belonged to this combined certification category (Peters-Stanley *et al.*, 2013) while a combined VCS and CCB standard captured about 47 per cent of the market share in 2014 (Peters-Stanley and Gonzalez, 2014).

The voluntary carbon market has become an important source of finance for forest conservation. It has been attractive due to its involvement of non-state actors and civil society, a small projects-focus and relatively low transaction costs, and also for allowing the trial of methodologies and instruments not yet approved in the regulatory framework (like the Reducing Emissions from Deforestation and Degradation REDD+ strategy) (Boyd *et al.*, 2009; Portaccio *et al.*, 2013). While forestry and land-use sector projects are barely represented in the compliance market, they are the most popular offset categories in the VCM (Hamilton *et al.*, 2008; Peters-Stanley *et al.*, 2013). According to the report on the State of Voluntary Carbon Markets 2013, forest and land use projects were the most popular in the VCM in 2012, accounting for about 34 per cent, and were followed by renewables (33 per cent), cookstoves (9 per cent), methane (9 per cent), energy efficiency (8 per cent) and gases (6 per cent) (Peters-Stanley *et al.*, 2013).

However, decisions on REDD+ in Warsaw during the Conference of Parties (COP) 19 in 2013 have raised questions about the future of forest-based projects in the VCM. The 'Warsaw Framework of REDD+', as it has come to be known, provides the grounds for a possible inclusion of REDD+ among the regulated/compliance forest carbon projects after 2015, when a new agreement on climate change is expected to be adopted during COP 21 in Paris. The effects of including

REDD+ in the compliance markets remain unknown, but are likely to restructure incentives and markets linked to forest carbon (Portaccio *et al.*, 2013).

Much will depend on the price of carbon in such markets. Without an international agreement, the price of carbon has dropped significantly in recent years. For instance, carbon price paid by voluntary buyers for forestry offsets fell from US$10.3/tCO2e in 2011 to US$7.6/tCO2e in 2012 (Peters-Stanley *et al.*, 2013) and further down to US$4.2/tCO2e in 2013 (Peters-Stanley and Gonzalez, 2014). The same trend is also observed in the compliance market. A recent report by a team of researchers from the World Bank Carbon Finance Unit argued that the price of primary CERs for most projects under the CDM went down from €3.86 in January 2013 to €0.34 in December 2013 (Ecofys, 2014). This has affected incentives to set up new projects significantly.

Table 2.1 summarizes the main differences and characteristics among the different carbon markets. As REDD+ is fast expanding as another mechanism giving opportunities for Africa to participate in the carbon market, a snapshot of its main features and architecture is compared with the CDM and the VCM.

TABLE 2.1 Main features of CDM, VCM and REDD+

Characteristics	Clean Development Mechanism (CDM)	Voluntary Carbon Market (VCM)	REDD+ (Reducing Emissions from Deforestation and Forest Degradation Plus)
Main Features	Kyoto Protocol targets can be met through the finance of clean technologies and low-carbon projects in developing countries.	Financing of clean technologies and low-carbon development projects through a voluntary system. Diverse methodologies allow a range of projects to sell credits, guaranteed by standards.	Focus on rewarding decreased rates of deforestation and forest degradation, as well as conservation, sustainable management and enhancement of carbon stocks. Aims to deliver social and environmental co-benefits to forest dwellers in developing countries.
Institutional framework	Decisions are taken by country Parties to the Kyoto Protocol through the CDM Board and CMPs. CERs are traded in national or regional emission trading schemes with a centralized government-regulated institutional structure.	Rules set by various VCM standards. Supply and demand is set by the market. VERs are primarily traded directly by projects or through retailers over-the-counter.	Institutional framework is being negotiated under the UNFCCC. National REDD+ readiness processes mostly follow UN-REDD Programme and FCPF guidelines, but other institutions intervene.

TABLE 2.1 continued

Characteristics	Clean Development Mechanism (CDM)	Voluntary Carbon Market (VCM)	REDD+ (Reducing Emissions from Deforestation and Forest Degradation Plus)
Financial mechanism	A market was created by legally binding emission reductions. Depends heavily on the Kyoto Protocol, but the market is contracting due to uncertainties over commitments.	A voluntary, private market was created linked to environmental and social responsibility aims. Improved methodologies, higher credibility and increased public interest in offsets are enhancing the market, although prices are low.	Credits currently do not contribute to countries' emission targets. Finance from multilateral and bilateral funds. REDD+ projects are growing steadily under the VCM. Finance for national programmes is slow-moving.
Activities	Afforestation and reforestation (A/R) only.	A/R, improved forest management, REDD+, and agroforestry.	REDD+ mitigation areas, plus activities to build capacities in developing countries.
Project cycle	Set by the CDM according to rules and modalities agreed under the UNFCCC: (1) project design (PD); (2) national approval; (3) validation of PD; (4) registration; (5) monitoring; (6) verification of emissions; (7) CER issue.	Set by multiple standards: (1) choose a methodology; (2) project description (PD) document; (3) validation of PD; (4) monitoring; (5) validation of emission reduction; (6) registration and VER issue.	National-level REDD+ readiness process is outlined by the UNFCCC and made operational by UN-REDD Programme and FCPF, centred around national REDD strategy.
Main actors	Project developers, CDM Policy Board, buyers, brokers and retailers, CMP, DOEs, DNA, consultants, Annex 1 governments running emission trading schemes.	Project promoters, standard certifying agencies, validation/verification bodies, buyers, brokers and retailers, consultants, 'carbon cowboys', Annex 1 governments running emission trading schemes.	Multilateral funds, bilateral funds, development agencies, international development banks, national development banks, UN agencies (UNEP, UNDP, FAO, UN-REDD), carbon consultancies, national governments.

Characteristics	Clean Development Mechanism (CDM)	Voluntary Carbon Market (VCM)	REDD+ (Reducing Emissions from Deforestation and Forest Degradation Plus)
Safeguards/ social impact	Sustainable development dividend defined and approved by designated national authority. Criteria are not defined by the CDM Board or UNFCCC but by the host government.	Criteria vary among VCM standards. Some focus just on measuring carbon sequestration, others verify community involvement and social and environmental co-benefits.	Social risks are mitigated and benefits supposed to be ensured through the establishment of international social and environmental safeguards information systems by the UNFCCC, UN-REDD Programme and FCPF.
Additionality	Measured against a baseline scenario based on current land uses or historical uses since 1990. Methodologies for defining baselines are set by the CDM.	Varies among approved methodologies. It may be historical or projected. Proponents may propose new methodologies for defining baselines.	Each country will develop their own methodology using IPCC guidelines.
Leakage	Risk of leakage must be calculated and emissions subtracted from removals claimed.	Risk of leakage must be calculated and emissions subtracted from removals claimed.	Leakage is controlled by measuring, reporting and verifying emissions at the national level.
Permanence	tCERs and lCERs are issued and are validated over 7–30 years, after which they must be replaced by permanent CERs, tCERs or lCERs.	VERs are issued over 20–100 years. A stock of untradeable credits is kept by the registry to compensate for non-permanence.	Still under negotiation at the national. At the subnational level, projects often follow rules set by voluntary standards.

Source: Adapted from Fong-Cisneros (2012)

Africa in forest carbon markets

The previous sections have described the various institutional and market mechanisms underpinning carbon forestry projects. In this section, we give an overview of the participation of African countries in compliance and voluntary markets, before turning our attention to discussing the emerging REDD+ mechanism, Africa's involvement in it and the flows of funding towards the continent.

Africa in the CDM

The CDM was the main mechanism giving Africa the opportunity to participate in the three mechanisms created by the Kyoto Protocol. But Africa is the most underrepresented geographical region. By the end of 2013, there were 161 countries involved in the CDM (UNFCCC, 2013) but only 2.4 per cent of all CDM projects are located in Africa. Africa has thus not benefitted yet from the CDM, in spite of the claimed benefits (Boyd *et al.*, 2007; Jindal *et al.*, 2008).

This has been attributed to several key factors, including lack of local expertise, the complex technical requirements of CDM and the general low emission baselines resulting from Africa's underdevelopment (Boyd *et al.*, 2007). By their design, the compliance carbon market favours projects that can deliver high emissions at the lowest cost with the lowest perceived risk (Linacre *et al.*, 2011). However, such favourable business conditions are often best found not in Africa but in middle-income countries such as China, India, Brazil, South Korea and Mexico, the largest supplier of CDM credits (Fong Cisneros, 2012). Other reasons limiting the success of the CDM in Africa include high transaction costs, particularly for projects with multiple, small emissions sources, the restrictions of land based sequestration and the focus of African governments on aid donors rather than the private sector (Portaccio *et al.*, 2013).

Voluntary carbon markets

Unlike the difficulties of the CDM in Africa, voluntary carbon market projects have seen a far greater degree of uptake of forest-based projects. This is partly because of the flexibility and less stringent rules of the market. Many of the CDM interventions increasingly focus on non-land-based projects (i.e. energy, biomass, landfill etc.) rather than land-based projects, and potentially leave out a significant number of rural populations (Benessaiah, 2012). The VCM has therefore remained a key source of carbon finance for forest conservation activities (Diaz *et al.*, 2011; Peters-Stanley *et al.*, 2013). Diaz *et al.* (2011) argue that just a little over 5 per cent of forest-sourced carbon offsets are traded through the compliance market, with the remaining 95 per cent being traded through the voluntary market.

Forest carbon projects therefore seem to be performing quite well in the VCM, particularly for REDD-type projects. According to Peters-Stanley *et al.* (2011), of an estimated 47 per cent of the credits produced by forest carbon projects and sold in the voluntary carbon market, about 29 per cent came from REDD-type initiatives. In 2013, REDD+-type interventions still dominated the voluntary market with about 38 per cent of the market share, ahead of clean stoves projects (24 per cent) and wind-based projects (7 per cent) (Peters-Stanley and Gonzalez, 2014).

Yet, comparatively, African contributions to the overall market remain small. Latin America is the largest supplier of carbon credits and in 2010 Latin America supplied about 58 per cent (largely REDD+ projects) of the total global primary market volume, with North America and Asia supplying 17 per cent and 15 per cent respectively (mostly from REDD+ and A/R projects). Africa only supplied about 8 per cent of the total volume (largely from REDD+ projects) (Diaz *et al.*, 2011).

In contrast with the low participation of Africa in both the compliance and the voluntary markets, the continent has nevertheless become one of the key destinations for testing the idea of REDD+. For instance, 44 developing countries participated in the World Bank-sponsored Forest Carbon Partnership Facility (FCPF) as of March 2014, with 17 from Africa, 16 from Latin America and 11 from the Asia-Pacific region.

Africa and the emerging REDD+ mechanism

The restriction of the CDM to only A/R schemes and subsequent exclusion of emission reductions resulting from avoided deforestation, avoided forest degradation and conservation, meant that sustainable management of forests and enhancement of forests became a very topical issue after the Kyoto Protocol was formalized (Angelsen, 2008). In 2005, the issue of reducing emissions from deforestation and degradation (REDD+) was raised on the floor of the UNFCCC for consideration in a post-Kyoto agreement. In 2007, REDD+ was officially agreed to be one of the mitigation policies for consideration in a future agreement during COP 13 in Indonesia. Developing countries were subsequently encouraged to explore REDD+ demonstrations and actions while developed countries were called upon to mobilize resources to support these efforts. In response to the COP 13 decision, numerous initiatives have been launched to develop national REDD+ programmes and sub-national REDD+ projects, now known as 'readiness' interventions (Angelsen *et al.*, 2012). COP 16 in Mexico outlined a sequence of three broad phases to develop a REDD+ mechanism under the climate change convention. The sequence starts with basic capacity building and development of strategies and action plans (phase 1), followed by the implementation of national strategies and results-based demonstration activities (phase 2) to eventual fully measured, reported and verified emission reductions (phase 3). In 2013, the UNFCCC completed negotiations on REDD+ after a series of controversies on areas such as the scope, funding, methodology for measuring emission reductions and safeguards during COP 19 in Warsaw (UNFCCC, 2014). REDD+ is expected to become part of the post-2012 Kyoto regime, although the final decision would depend on COP 21 in Paris in 2015.

Amongst the most important international initiatives supporting national REDD+ readiness are the FCPF, launched in December 2007 and the UN-REDD Programme launched in 2008. The UN-REDD Programme is a partnership between the United Nations Food and Agriculture Organization (FAO), the United Nations Development Programme (UNDP) and the United Nations Environment Programme (UNEP). Of course, there are several other bilateral agreements on REDD+ projects ongoing. The FCPF is a global partnership of developed and developing countries which has the World Bank as trustee and main delivery agency. As of March 2014 there were 44 countries engaged in the FCPF Readiness Fund (Forest Carbon Partnership Facility, 2014) and 50 partner countries engaged in the UN-REDD (UN-REDD, 2014). Seventeen of the 44 participating countries in the FCPF are located in Africa, while about 20 partner countries of the UN-REDD programme can be located on the continent. Between 2008 and 2010, some US$7 billion was estimated to have been spent in supporting REDD+ early actions across the world (Simula, 2010). In 2010 REDD+ projects provided 67 per cent of credits sold in the primary market (Diaz et al., 2011). From 2009 to 2010 the share of REDD credits in the carbon market rose by 500 per cent (Linacre et al., 2011).

Unlike the CDM that witnessed the participation of relatively few African countries, there are more than 20 African countries currently engaged in the REDD+ readiness at the national level under the FCPF and/or the UN-REDD programme, including Kenya, Tanzania, Ghana, Zambia and Zimbabwe (see various chapters in this book).

While there is growing bilateral and multilateral financial support for REDD+ activities over the years, this is not evenly distributed among geographical regions (Simula, 2010; Fong Cisneros, 2012). Latin America and the Caribbean receive about 44.7 per cent of funding, while Africa receives 17.6 per cent of approved finance at US$235 million (Climate Funds Update, 2012 cited in Fong Cisneros, 2012). Guyana, Brazil, Indonesia and Mexico and then DRC and Tanzania are by far the highest recipients of approved finance from climate funds. Within Africa, some countries are leaders in terms of capturing funds for REDD+, whereas others lag behind. Consequently, finance for countries has ranged from about US$0.2 million (as in the case of Ethiopia, Kenya, Libya, Uganda) to US$52.2 million (DRC) (Fong Cisneros, 2012).

While REDD+ is expected to bring benefits for climate change mitigation, sustainable development and also for forest-fringe people through, for example, poverty reduction from cash transfer, the creation of employment or the protection of local environment (Brown et al., 2008; UNFCCC, 2011; Parrotta et al., 2012), challenges with REDD+ programmes in practice are increasingly documented. Negative impacts include displacing forest-dependent communities, restricting access for people to the use of forest resources and limiting participation of forest-dependent communities (Hall, 2007, 2008; Chhatre et al., 2012; Ribot and Larson, 2012). In an attempt to minimise these negative consequences, the UNFCCC has adopted decisions on development and provision of information on safeguards as a key requirement for ensuring that REDD+ interventions 'do no harm' but respect rights and provide equitable benefit-sharing frameworks in favour of local communities (UNFCCC, 2011). But safeguards, and the ideals they espouse, are becoming buzzwords with embedded confusion and conflation (Arhin, 2014). This weakens the hopes and the potentials of safeguards to guard social justice ideals under REDD+.

As discussed in the opening chapter of this book, the extent to which national REDD+ strategies and policies – and carbon forestry interventions more generally

– will be able to deliver equitable, efficient and effective outcomes now occupy central positions of the REDD+ debate (Peskett *et al.*, 2011; Visseren-Hamakers *et al.*, 2012). One question which increasingly is asked is: Will REDD+ succeed and will it be effective in addressing social justice and livelihood needs of poor forest-based communities? In the next section, we turn our attention to this question and sketch out some entry points that can facilitate carbon projects, and particularly REDD+, to become one of the pathways for sustainability and a mechanism for enhancing livelihood needs and benefits for local communities.

Enhancing equity in forestry carbon projects

Can forestry carbon projects, with their emphasis on market efficiency, provide a pathway to sustainability that also meets equity principles and livelihood priorities of small farmers and forest dwellers in rural African settings and beyond? The answer depends largely on the governance of market and project processes, the types of standards and associated measurement and verification requirements, the institutional arrangements deployed and how issues of equity, social justice and livelihood improvements are prioritized in the design and implementation of carbon forestry projects.

For African nations and other developing countries there remains a longstanding challenge of reconciling the urgency to reduce poverty and to protect natural resources while at the same time meeting an increased market demand for forest products. The case of carbon projects as advanced by the Kyoto Protocol was therefore very appealing to policy-makers as such projects were meant to address these apparently conflicting policy goals. But in the evolution of carbon forestry projects, and the associated institutional, policy and market arrangements that support them, these social justice issues have not been a major focus. It is only now, when projects are being rolled out on the ground, that these issues have become live. To date, economic and technical matters have dominated the discussion on carbon projects, with less attention being paid to issues of equity and sustainable development (Boyd, 2002; Brown and Corbera, 2003; Corbera *et al.*, 2007). Indeed, while the UNFCCC reports that some one billion tonnes of CO_2 have been reduced through carbon projects since 2004 (UNFCCC, 2013), a mass body of research points out that most carbon projects have fallen short of their equity and local development objectives (Brown *et al.*, 2004; May *et al.*, 2004; Sirohi, 2007). With forest carbon projects occurring in lived-in landscapes, where conflicts over land and resources are common, and where poverty is widespread and livelihoods are complex and fragile, placing issues of equity, social justice and sustainable development at the heart of projects offers potent pathways to promote carbon benefits for local communities.

The inclusion of small-scale projects in the final decision on LULUCF under Article 12 (Decision 19/CP.9) of the Kyoto Protocol sought to broaden the scope of beneficiaries of carbon projects in favour of low income communities (Boyd *et al.*, 2004; Boyd *et al.*, 2007). This has brought to the fore the need to channel additional financial resources and benefits from projects to communities. Yet, in Africa, explicit benefit-sharing arrangements in favour of communities are, to date, almost non-existent. This remains a challenge for forest carbon projects to achieve desired objectives of poverty reduction and sustainable development. If carbon projects are to provide a pathway for sustainability and social justice, then benefit-sharing frameworks that recognize communities as key beneficiaries should be central in project design and implementation.

Another key issue thrown up by the implementation of carbon forestry projects in Africa is the need to clarify tenure and use rights, not only for land but also for carbon pools (Larson and Petkova, 2011). In many parts of Africa, forest owners and users are in many cases not the same and this has implications for forest management. Within the current REDD+ debate, unclear land tenure has raised questions such as: Who owns the emission reductions and therefore who should be compensated for any reductions in emissions, and how are these payments to be distributed across owners and users? With billions of people living in or near forests and making use of forest lands and resources but having no or few secure rights nor tenure over these lands and resources, the need to clarify and ensure adequate rights for local communities and marginalized groups has been highlighted as key to achieving social justice.

Conclusion

This chapter has sought to provide an overview of the different types of forest carbon mechanisms in which African projects and programmes participate (i.e. CDM, VCM and the emerging REDD+ programme). Forestry related CDM projects globally are very small relative to other types of CDM projects, and Africa has been hugely underrepresented as producers of CERs from A/R projects. But unlike the difficulties of the CDM in Africa, voluntary carbon market projects have seen a relatively greater uptake of forestry related projects, although Africa still lags behind Latin America and Asia. The VCM has remained a key source of carbon finance for forest conservation activities, particularly REDD+ type projects. National governments have had profound roles in the implementation of CDM projects, and now REDD+, while a variety of private entities and charities lead in the activities of the VCM.

Whilst sustainable development and improvement in livelihoods have been part of the narratives around which carbon projects are justified, they have fallen short of delivering these benefits in practice. Repositioning forest carbon projects to benefit local communities will involve taking issues of social justice and equity seriously, including a focus on designing equitable benefit sharing arrangements and addressing land tenure issues.

Today, forestry carbon projects are vital components of current thinking on global environmental governance and management of natural resources. Forest carbon projects are developing in a very complex manner. They are structuring institutional frameworks and are increasingly dominated by international actors.

National governments, the private sector, the NGO sector, philanthropists and researchers all hold different interests, roles and levels of participation and power. But, crucially, local communities have had only a marginal influence and position in the whole carbon policy web (Fong Cisneros, 2012). The high levels of technical, financial and procedural complexity set by multiple institutions around forest carbon are mostly inaccessible for local people, and not just marginalizes but excludes communities from meaningfully engaging in the process. As the chapters in this book show, this can represent a key source of conflict and resistance to projects. It is time for carbon projects to place issues of social justice, equity, livelihood benefits and sustainable development ideals at the heart and centre of their implementation.

Note

1   See list of acronyms and abbreviations for full details.

3

CLIMATE EMERGENCY, CARBON CAPTURE AND COERCIVE CONSERVATION ON MT. KILIMANJARO

Martin Kijazi

Introduction

This chapter examines the politics of climate and carbon forests on Mt. Kilimanjaro in Tanzania. Global concerns around climate change and the emergence of forest-based carbon sequestration projects has raised the profile of forests in national and local politics. The result has been a whole array of initiatives on and around Mt. Kilimanjaro. The chapter examines the policy discourses and political and institutional practices that have emerged, as well as local perspectives and forms of resistance.

The declaring of a climate 'emergency' has resulted in a number of initiatives, mostly top-down and exclusionary, as the carbon resource is captured and coercive conservation returns to Mt. Kilimanjaro. For example, the Regional Commissioner's (RC) office has declared a ban on tree-cutting on public and private lands in Kilimanjaro. The Regional Administrative Office also houses a United Nations Development Programme (UNDP) and Global Environmental Facility (GEF) 'Sustainable Land Management' (SLM) project which is prospecting on 'scaling-up' carbon-financing, as a response to assumed local forest 'degradation' as the major contributor to climate change. Officials also blame villagers' practices, ignoring the role of many complex and dynamic political, economic and ecological drivers of change. Conversely, this policy narrative unrealistically views local forests as a panacea through which all climate, carbon emissions and environmental woes will be erased. Thus, central and regional government decision-makers have undertaken a series of interventions, including changes of laws to reclassify the forests of Mt. Kilimanjaro and the associated forest institutions. This reconfiguration of land access and authority has reversed earlier efforts to devolve forest management to local people. The changes have favoured centralization of forests and a renewed form of 'fortress conservation'. Currently, the state's coercive apparatus, combined with external climate mitigation project funds, is used to enforce this, keeping local people out. This reconfiguration is claimed to protect ecosystem integrity, yet there is little evidence to show that it is working.

Rather, as the chapter shows, syndicates of corruption involving the elite are continuing to extract valuable timber species. However, the local people are not simply sitting outside the fortress as spectators. Widespread forms of local resistance are also observed. The chapter thus highlights the often ignored or misinterpreted issue of local resistance, and argues that carbon forestry projects need to take seriously issues of justice, representation and accountability if they are to have any chance of success.

The carbon finance project focuses on carbon mitigation via energy efficiency or switching from wood to other sources (GEF, 2010). It is supposed to be registered through Clean Development Mechanisms (CDM), but has not completed this registration yet (Mutimba, 2013). Regarding benefiting or involving local populations, the project is not inclusive. The implementation has involved only 12 public and private institutions, none of which are affiliated with rural/village communities. Also, policy-wise and institutionally, the carbon finance project is interlinked with previous and current conservation interventions that now exclude local people via strict forest preservation also known as 'fortress conservation' (cf. Brockington, 2002). There are two ways in which the carbon finance project is linked to fortress conservation. First, the project comes after a forest centralization intervention that involved annexing forests of Mt. Kilimanjaro in the national park. This has led to highly contentious exclusions of local people from accessing resources from those forests. Instead, the forests are supposed to support a wider attempt at climate amelioration in Mt. Kilimanjaro, including carbon conservation. This is impacting negatively on the livelihoods of villagers, and has led to the array of resistance that I document in this chapter. Such exclusion of the broader population, resulting in local resistance, jeopardizes potential achievements of carbon mitigation of the carbon finance project. Secondly, as the carbon finance project includes only a limited number of institutions, the majority of rural people in Mt. Kilimanjaro are forced to conserve forests and trees on their farms via executive orders – particularly the tree-cutting ban issued by the Kilimanjaro Regional Office.

Institutionally, the fortress conservation and carbon finance interventions are interlinked. The fortress conservation interventions have been the initiatives of the central government agencies including the Tanzania National Parks Authority (TANAPA) and the Kilimanjaro Regional Office. The former is a parastatal organization. The latter is a deconcentrated authority of the central government. The former pioneered the annexation of Mt. Kilimanjaro forests in the park. The Regional Office is not only responsible for the current fortress conservation via the tree-cutting ban, but also for the implementation of the UNDP and GEF carbon finance project. While the UNDP and GEF are not directly implicated here, their own interventions have supported and/or worked alongside these fortress interventions. A project of UNDP and GEF in Kilimanjaro named the Community Management of Protected Areas Conservation Trust (COMPACT), funded the so-called 'threats on Mt. Kilimanjaro' study (Lambrechts *et al.*, 2002). This study was subsequently used to support TANAPA's discourse, which painted local people as destroyers of the forests. Hence it was successfully used in lobbying for the annexation of the forests in the park. This was confirmed by a UNDP/ COMPACT project official during an interview.[1] Also, given that the UNDP/ GEF carbon finance project is implemented by the Regional Office, it is entangled with the fortress conservation policies of the latter. While the carbon finance project, per se, is tied to possible future compensations of the participants, the broader fortress conservation interventions imposed on local populations are not tied to any current or promissory financial compensation scheme.

National context and carbon initiatives

Tanzania's mainland has a total of 33.428 million hectares (ha) of forests (37.8 per cent of the landmass). The deforestation rate is estimated to be around 1.2 per cent per annum (URT, 2013, p.xi). Key drivers of deforestation and forest degradation include high wood for energy uses, policy failures, argicultural expansion, urbanization and poverty (Nashanda, 2013). Tanzania is addressing these via a multi-sector approach (Nashanda, 2013; URT, 2013). Tanzania has also used its framework for sustainable forest management; participatory forest management (URT, 1998); its national forest programme; private sector involvement; Sector Reform Programmes; and the National Strategy for Growth and Reduction of Poverty (abbreviated in Kiswahili as MKUKUTA) (Nashanda, 2013).

But major challenges still abound, including weak financial capacity, coordination and governance. Tanzania has, thus, engaged with bilateral and multilateral climate and carbon forestry initiatives hoping to lessen these challenges. It has been involved in the United Nations Framework Convention on Climate Change (UNFCCC) processes since 2005; the Bali Action Plan (2007) and Decision 2/CP.13, which guided key components on Reduced Emmissions from Deforestation and Forest Degradation (REDD) readiness (Nashanda, 2013). Tanzania has developed REDD+ policies, including the National REDD+ Strategy (2013) and Action Plan. It has implemented nine REDD+ pilot projects. Some projects have provided incentives for people engaged through benefit distribution systems, where lessons and experiences have been documented for future policy development (Njaidi, 2014).[2] Other projects are developing the country's reference emission levels (REL). Others are developing modalities to address leakage and permanence, and establishing Monitoring, Reporting and Verification (MRV) systems (URT, 2013).

The Tanzanian–Norwegian Climate and Forest Initiative included developing REDD+ policies; training and research on climate change impacts, adaptation and mitigation (CCIAM); developing MRV capacities, establishing the National Carbon Monitoring Center (NCMC), and Piloting REDD+ (URT, 2013). The UN-REDD Tanzania Programme is implemented by the government of Tanzania with three UN agencies (FAO, UNDP and UNEP). It aims to (1) strengthen national governance framework and institutional capacities; (2) increase capacity for capturing REDD+ elements within National MRV; (3) improve capacity to manage REDD+ and provide forest ecosystem services at district and local levels and; (4) broaden stakeholder support for REDD+. The Clinton Climate Initiative deals with national carbon accounting. Tanzania–Finland-FAO – Technical and Financial support have developed the National Forest Resources Monitoring and Assessment approach (NAFORMA) (Nashanda, 2013).

The carbon finance initiative I examine in this chapter makes use of the post-Kyoto CDM, and the UNDP Millennium Development Goals (MDG) Carbon Initiative, and is linked in with the wider national REDD+ and climate mitigation policy initiatives described.

However, the new project interlocks with pathways of landscape change and political economy established by prior interventions. In Mt. Kilimanjaro, the more dramatic signals of climate change, including the estimated 80 per cent loss of glaciers, brought climate emergence and its associated politics to the fore. The last decade saw a strong interest to centralize protection of Mt. Kilimanjaro forests (cf. TANAPA, 2001; TANAPA and IRA-UD, 2002). This created a merger of 'green developmentalism' and 'fortress conservation' (Kijazi, forthcoming 2014). Green developmentalism is a supranational environmental and finance institutions' approach to regulate global flows of natural capital via eco-development funds and a new global discourse (McAfee, 1999), while fortress conservation seeks to preserve wildlife and their habitat through forceful exclusion of local people who have traditionally relied on the environment in question for their livelihoods (Brockington, 2002).

In Mt. Kilimanjaro, green developmentalism has been used to sensationalize triple-win solutions for the environment, economy and society. But the dominant approach has been fortress conservation. Notably influential frameworks have been the green development agenda of UNDP and GEF. These interlock with the Kilimanjaro National Park Authority (KINAPA) and Regional Commissioner's approach to bureaucratic control and coercion of local people.

This chapter examines the playing out of these initiatives in Mt. Kilimanjaro, with global and national discourses intersecting with local contexts, practices and resistance. As I will show, the particular landscape history – and past interventions to conserve the mountain's important ecosystems – have shaped the current intervention. This has given rise to a particular politics that draws on deeper memories and experiences. The failure of these fortress-style carbon interventions, and the increasing mobilization of local communities against them, shows how projects designed in the abstract encounter diverse politics on the ground, shaped by long histories.

Landscape context, ecology and environmental history

Mt. Kilimanjaro is located in north-east Tanzania. It towers at 5895 m.a.s.l. over the surrounding plains. Given its height, Mt. Kilimanjaro plays an important climatic role. It regulates temperature and hydrology – on its slopes and on the surrounding plains – through the interception of wet monsoon winds from the Indian Ocean. The slopes receive heavy rainfall, and the peak altitudes are covered by snow or glaciers. Mt. Kilimanjaro has a crucial water catchment value given its extensive forests and high rainfall. About 96 per cent of water flowing from Mt. Kilimanjaro originates from the forests. The water is used domestically, industrially, for irrigation and in fishing schemes. Hydropower generation from its distributaries contributes about 20 per cent of Tanzania's power (Lambrechts *et al.*, 2002, p.7).

Mt. Kilimanjaro's diverse ecosystems host diverse flora and fauna: an estimated 900 forest plant species; 2,500 species for the entire mountain, and about 140 mammal species (Lambrechts *et al.*, 2002, p.9). It is the most populous tourist destination in Tanzania, attracting more than 35,000 annual climbers, with earnings from the total in-country tourists' expenditure at around US$50 million (equivalent to 80 billion Tanzania shillings) (*Tanzania Daily News*, 2013). The tourists numbers peaked at 57,456 in 2011/2012 (TANAPA Press Statement, 5 March 2014, p.1). Its forests provide various economic products, including timber, honey, fuelwood, nuts, fruits, seeds, poles for construction, etc., and commercially important timber plantations.

Abundant rainfall, forests and fertile volcanic soils have attracted heavy human settlement. Recorded continuous human settlement is at least the last 2,000 years. The current inhabitants (Chagga people) arrived in successive waves of multi-ethnic migrations starting at least five or six centuries ago (Odner, 1971). They merged to become Chagga in the eighteenth century (Greenberg, 1966). They transformed portions of the original forest into an agroforestry system called the Chagga home-gardens (*vihamba pl.*, *kihamba sing.*) that harbour multi-layered strata of herbaceous crops, e.g. taro yams, and perennial crops, e.g. coffee and banana, and trees (Kijazi, 2007). The Chagga also created a canal and furrow irrigation system to deliver water from the forest into their *vihamba* (Johnston, 1886).

Many local Chagga and their stall-fed livestock still inhabit the nucleus of their *vihamba*. Their economy and livelihoods are intricately linked to the forests. A network of canals and furrows feeds the *vihamba* with water from the forest. Villagers collect firewood and livestock fodder there. The animals provide meat, milk and cash for the homestead, plus manure for crops. The multipurpose trees on *vihamba* provide fuelwood, fencing, fodder, mulch, bee forage and shade for crop trees, particularly coffee (Kijazi, 2007).

The mountain also contains important cultural sites of the Chagga people; for example, the Kifinika cone that is used for traditional ceremonies (TANAPA, 2001). The forests harbour many plants with traditional, cultural and magical values. An ethno-botanical study showed that the Chagga make diverse uses of their plant environment, including foraging for household and agricultural purposes and ethno-medicinal applications. One hundred and seventy-two plant species with medicinal purposes were recorded in the Old Moshi area (Lambrechts *et al.*, 2002, p.10).

Despite this long history of customary resource use, new forms of fortress conservation precipitated by the carbon rush continues a legacy of state land alienation from colonial times to the present. Large tracts of agricultural and grazing lowlands below the *vihamba* were alienated by the German colonial government for settler farm estates. Simultaneously, upper-slope forests above the *vihamba* were alienated for reservation. The 1904 Forest Conservation Ordinance during German rule declared Mt. Kilimanjaro a game reserve. As Neumann wrote, the effect of German forest laws on existing African access and use was direct and immediate:

> Under Teutonic discipline (which included corporal punishment and confinement in chains), all African settlement, cultivation, burning, and grazing was outlawed in designated forest reserves.
>
> *Neumann, 1998, p.99.*

Mt. Kilimanjaro was gazetted as a forest reserve in 1921 by the new British colonial government. Given local people's hardships, local chiefs negotiated with the British rulers. A buffer zone called the 'half-mile-forest' was demarcated on the edges of the Mt. Kilimanjaro forest reserve in 1941, to sustain local communities' livelihoods. It was managed as a 'social forest' by the Chagga Council for 20 years. From 1962, following independence and abolition of the native authorities in Tanganyika, the half-mile-forest was managed by district councils. From 1972 it was managed by the central government, following the abolition of local governments, until 1987 when it was returned to re-established district councils (Kivumbi and Newmark, 1991). The core forest reserve (107,828 ha), excluding the half-mile-forest, was managed as a state/catchment forest by the South Kilimanjaro Catchment Forest Authority. The area (75,353 ha) above the forest line (2,700 metres) was reclassified as a national park in 1973 by the Tanzanian Government. The park was inscribed on the World Heritage list in 1987 (Lambrechts *et al.*, 2002). Promoting world heritage site status was subsequently used to justify the recent annexation of the forest reserve to the park (TANAPA, 2001) in 2005. Then, the local government half-mile-forest was also alienated and annexed to the park. Subsequently, a tree-cut ban was also declared on public and private lands in the Kilimanjaro region, with the intentions of 'stabilizing the climate and reversing environmental degradation in Kilimanjaro' (Kilimanjaro Regional Commissioner, 2013).[3]

The annexation of forests was based on a promise to protect the ecological integrity of Mt. Kilimanjaro, and to reverse and stabilize the current negative ecological trends (TANAPA, 2001). However, these forests also harbour highly valuable commercial timber species, e.g. Camphor wood (*Ocotea usambarensis*), African pencil wood (*Juniperus procera*) and Podo (*Podocarpus mylanjianus*), which has attracted elite interests in illegal logging, fuelled by corruption (MNRT, 2003). While the park authority promised to curb illegal logging, almost a decade later illegal logging still abounds. Local key informants and investigative journalists allege involvement of park officials (cf. Kijazi, forthcoming 2014). The presumed effectiveness of fortress conservation to protect ecological integrity seems futile, and the role of carbon finance seems to be having little effect on the main cause of forest degradation.

Policy narratives: Creating the justification for forest carbon interventions

Before the park expansion, two competing policy narratives were promoted by two rival state institutions in the Ministry of Natural Resources *viz*. the Forestry and Bee-Keeping Division and the Tanzania National Parks Authority. The former and its interlocutor, the South Kilimanjaro Catchment Authority (collectively, 'the forestry authority', henceforth) argued that inadequate inclusion of villagers in forestry governance was to blame for deforestation and forest degradation. Therefore, it promoted participatory forest management as per the forestry policy (URT, 1998):

> They [South Kilimanjaro Catchment Forest Authority] stipulate collaboration with local communities adjacent to catchment forests through Joint Forest Management Agreements in order to guarantee sustainable management and reduction of forest management costs through sharing of benefits and responsibilities [and] to facilitate community based conservation and devolution of power to local and district authorities.
>
> *Akitanda, 2002, p.3.*

The ministry should look into the possibility of capturing greater forest gate receipts from the Kilimanjaro catchment forest reserve for the purpose of developing Joint Forest Management in the area. (MNRT, 2003, p.iv)

The Tanzania National Park Authority and its interlocutor, the Kilimanjaro National Park Authority (collectively, 'the park authority', henceforth), presented a counter narrative vilifying Joint Forest Management Agreements (JFMAs) with villagers. Instead, it proposed annexing the forest to the park, and heightening park ranger patrols (TANAPA, 2001; TANAPA and IRA-UD, 2002), establishing a form of 'fortress conservation'. This narrative has colonial origins (Neumann, 1998). It propagates a myth of unpopulated, unspoiled wildernesses as the protected area park ideal, and it presumes incompatibility of native Africans' land use with conservation (Brockington *et al.*, 2008). The park authority proposal to annex Mt. Kilimanjaro forests states:

> If annexation of the catchment forests will not be done, local people will continue to have uninterrupted access to the catchment to obtain forest products. This will produce a false sense of security in the short term […] In the longer term the forest and its catchment values will decline. This will have a local, regional and international damage to the world heritage site. E.g. Tourism activities may be impaired.
>
> *TANAPA, 2001, p.36.*

Despite their historical forest rights, villagers were vilified as trespassers and invaders (cf. TANAPA, 2001; TANAPA and IRA-UD, 2003). Tourism was glamorized, despite environmental effects caused by tourist over-crowding, vehicle pollution and immense garbage. If local people are fenced out, the proposal remarks:

> Tourism and related business will improve significantly at the international, national, and at the local levels respectively as Mt. Kilimanjaro will continue to attract more tourists hence more income generation.
>
> *TANAPA, 2001, p.37.*

Another study commissioned by the park authority dismisses options for local use, and asserts:

> Forest Policy advocates for Joint Forest Management. Such initiative has been introduced for Mt. Kilimanjaro Forest Reserve […] This concept works on the basis that once the communities living adjacent to the forest reserve realize some benefits from the forest then it is an incentive for them to actively engage in the management [but] their demand for forest product is more commercial rather than subsistence such as logging for extraction of high quality timbers […] this resource extraction is not sustainable. It is on this basis that the proposal of annexing this forest reserve as part of Mt. Kilimanjaro National Park is valid.
>
> *TANAPA and IRA-UD, 2002, pp.15–16.*

Here, to justify villagers' exclusion, the study strategically conflates their forest needs with 'illegal logging', ignoring that the latter is an elite activity; while villagers' uses primarily cater for livelihoods (cf. MNRT, 2003). To complete this vilification, villagers' activities were also blamed for 'melting' Mt. Kilimanjaro glaciers (TANAPA, 2001, p.22, p.27):

> There is continued degradation […] particularly in the forest reserve and half mile forest strip. This has led to changes in ecological and climatic patterns. (p.22) […] This rapid retreat of Mt. Kilimanjaro ice fields is suggested to be partly attributed to climatic changes globally (warming) but most of it is due to acidification, which is mainly due to clearance of the vegetation on Mt. Kilimanjaro feet. (p.27)

Exaggerating impacts of local activities and their supposed negative impacts on global climate change is not supported by existing extensive research, but the narrative served particular interests. The parks authority recommends that 'annexing the catchment forests to KINAPA will control the current rate of destruction stop, it is likely that the climatical conditions of the area will improve' (TANAPA, 2001, p.34).

Here, climate emergency sensationalism paved a discursive path for fortress conservation. At the time, the forestry authority had the legal mandate to govern Mt. Kilimanjaro catchment forests, but it was fraught with financial difficulties, with heavy donor-dependence (cf. Akitanda, 2002). Its rival, the Kilimanjaro National Park authority, is a cash machine. It pays its administrative costs requiring no subsidiaries; rather, it pays corporate tax to central government (cf. Kijazi, forthcoming 2014). Thus, the park authority also argued that its greater earnings would fund more effective anti-poaching patrols to protect the integrity of the World Natural Heritage Site (WNHS) status of Mt. Kilimanjaro (TANAPA, 2001).

The WNHS advocacy created a common interest with the UNDP/GEF community management of protected areas project (COMPACT). COMPACT's primary goal was to 'add value to existing biodiversity conservation programmes in and around WNHS through community-level approaches' (COMPACT, n.d., p.1). Despite the 'community level approaches' rhetoric, COMPACT's primary concern was preserving biodiversity and WNHS. To achieve its objective to 'reduce threats to biodiversity attributed to local communities' (COMPACT, n.d., p.1), COMPACT supported the park authority's forest annexation advocacy and lobby. COMPACT commissioned a study on 'threats to Mt. Kilimanjaro'. The study mapped human uses inside the forest (Lambrechts et al., 2002). But the snapshot mapping lacked historical-social-cultural contexts, including local forest claims and rights. Mapping all human uses as red dots labelled as 'THREATS' on forest maps mixed livelihood uses with environmentally damaging elite uses (e.g. illegal logging). These visually powerful maps were used as a discursive tool for lobbying the central government to stop these 'THREATS' by legislating the park expansion. The maps still hang on walls in government and NGOs' offices, reminding any official or visitor of local threats, and reinforcing the fortress conservation narrative. While the park authority pushed villagers out of the forest by force, COMPACT pulled them out of the forests using incentives. Thus, in this instance, green developmentalism interlocked with fortress conservation to exclude villagers from forest benefits and decisions on its use (cf. Kijazi, forthcoming 2014).

The annexation was legislated by the Tanzania national assembly in 2005. Viewing local people as 'threats' had set a discursive precedent that was later exploited by the Kilimanjaro RC's office. Justifying the subsequent alienation of the local government half-mile-forest in 2009, and entrusting it also to the park authority, the RC remarked:

> In order to protect against serious threats to the climate of Mt. Kilimanjaro, we had to intervene further. When I moved to Kilimanjaro, I was first shocked to notice how hot it had become. One day it reached 40° C. I heard about a lot of forest destructions. Then I had an opportunity to fly over the Mt. Kilimanjaro forests with the Tanzania National Parks director. We noticed the horrific state of the forests. I called several meetings. So, we took action. We needed to stop the haphazard tree felling in Kilimanjaro that is melting the glaciers, drying water we badly need for our economy, and making Kilimanjaro hot like a desert.
>
> *Kilimanjaro Regional Commissioner, 2013.*[4]

This account reflected a subjective sensationalism of short-term patterns, and an articulation of an unfounded but well-entrenched narrative, rather than an objective analysis of circumstances and trends. The argument failed to examine the underlying causes of forest degradation. Illegal logging was being carried out by informally sanctioned local elites, and could not be attributed to local people. Yet, villagers are vilified. The RC admitted observing illegal logging even in the annexed reserve, seven years after the annexation. The presumed superiority of the park expansion and fortress conservation should therefore have been questioned. Instead, the park authority received the alienated local government forest. Subsequently, in 2012, the RC consulted the Regional Consultative Committee. Following the meeting's resolutions, the RC's office ordered the tree-cutting ban on public and private lands in the Kilimanjaro region with the exception of central government industrial plantations. The process of re-centralizing the 'climate forests' of Mt. Kilimanjaro continued. The recent carbon finance project discussed in the next section, therefore, has not started on a blank slate. It has been deeply influenced by historical experiences and dominant narratives around the causes and consequences of forest degradation.

The Mt. Kilimanjaro carbon finance project

The carbon finance project is a component of the 'Project for Reducing Land Degradation on the Highlands of the Kilimanjaro Region', also called 'Sustainable Land Management (SLM) – Kilimanjaro'. The four-year project (September 2010–December 2014) is a partnership between the United Nations Development Programme – Tanzania (UNDP-Tanzania) and the Office of the Vice President-Department of Environment (VPO-Environment). According to the consultant Camco, the Energy for Sustainable Development consulting firm that is developing the carbon-finance project:

> Through the project, the government is seeking to engage with public institutions to reduce emissions from both the inefficient use of biomass energy in a triple-win situation that cleans the environment, reduces deforestation and provides avenues for income/revenue generation from carbon market. The project will focus on sustainable harvesting, efficient conversion of wood to energy, and energy saving techniques.
>
> *Mutimba and Kibulo, 2013, p.5.*

The project is co-funded by the Government of Tanzania and the Global Environment Facility through the UNDP. Nationally, the implementing agency is UNDP-Tanzania. The VPO-Environment is the implementing partner, with others: the Ministry of Energy and Mineral Development; Ministry of Agriculture; and Kilimanjaro Regional Government. In Kilimanjaro, the Office of the Regional Administrative Secretary (RAS) is the key implemeter, working with seven administrative (district) councils, namely Hai, Siha, Moshi Rural, Moshi

Municipality, Rombo, Mwanga and Same. A Project Steering Committee oversees the activities of Regional Technical Team and District Facilitation Teams (RAS-Kilimanjaro, 2013).

The stated goal is:

> [T]o design an energy improvement strategy for the Kilimanjaro region whose implementation will lead to emissions reduction linked to a carbon credit earning scheme, preferably through the UNDP MDG Carbon Initiative and other carbon market schemes.
>
> *Mutimba and Kibulo, 2013, p.6.*

Camco has shortlisted institutions that may participate in the project based on the following criteria: (1) willingness to share information and to participate; (2) having a range of energy efficiency measures, e.g. improved stoves, woodlots, biogas, solar, etc.; (3) land available for deploying project technologies and interventions; and (4) plans for improving energy efficiency and reducing fuelwood consumption. The qualification screening was based on questionnaires used in a survey of 204 institutions. Selected institutions were then visited to: (1) assess fuel consumption – quantity and cost; (2) obtain copies of invoices, receipts, fuel purchase records and electricity bills; (3) record the number and models of stoves used for cooking and heating; (4) assess the physical condition of these technologies, and record operation and maintenance arrangements; (5) view and record other energy efficiency initiatives in place; and (6) obtain assurance from institutions that they are still interested in participating (Mutimba and Kibulo, 2013). By the end of 2013, 12 institutions had been shortlisted for participation in the carbon finance project. These include six secondary schools, two colleges, a university, a hospital, a policy academy and a prison.

The selected institutions are estimated to have a population of about 12,256 individuals (Mutimba, 2013, p.47). The Kilimanjaro region had an estimated population of 1,635,870 in 2010, of which 90 per cent of both rural and urban communities use firewood and charcoal for cooking (Mutimba, 2013, p.16). Thus, the estimated 12,256 individuals involved in the carbon finance project represent a tiny fraction of the population. The remainder of the Kilimanjaro residents are coerced to switch from their reliance on local forests for energy and other livelihood uses by the fortress interventions that preceded the carbon finance project, including banned or restricted access to surrounding forests and the tree-cut ban on their private farms.

The project focuses on introducing new energy technologies to reduce deforestation, combined with some new tree planting. Due to the conflicts over land and tree tenure in Kilimanjaro, the project evades confronting these problems by strategically engaging with institutions that are not entangled. Government institutions, including schools and others, have been targeted instead.

However, the overall ambitions of the project are restricted. A project official mentioned that the greatest challenge of the project is land shortage: 'So far, none of the land offered for woodlots is more than 5 ha. Only institutions selected for carbon financing have offered larger plots' (Anon., 2014).[5] But if the outcomes (e.g. woodlots) are to be scaled-up to the broader population, it is hard to imagine success without addressing the land, forest and tree tenure/rights issues in the area.

Revaluing forests for carbon

The project expects 'avoided deforestation' due to an energy switch from wood-intensive sources to alternative sources, or wood saving technologies. The project targets at least 25 per cent recovery in highly degraded patches as measured by regeneration (recruitment) and improvements in species index for forests/woodlands (GEF, 2010, p.33). One success indicator is a reduction in the rates of deforestation, from the project's baseline estimated deforestation rate of 6 per cent/year. This figure, which is provided in the project proposal (GEF, 2010, p.33), is significantly higher than the national deforestation figure of 1.16 per cent/year (URT, 2013, p.xi). Yet it is neither validated in the project proposal, nor current project implementation documents examined, nor by project interview officials interviewed including the Regional Natural Resource Officer in the Regional Administrative office where the project is being implemented.

Participants will have to demonstrate their contribution to avoided deforestation in order to obtain their Certified Emission Reductions (CERs). This requires completing the Clean Development Mechanisms (CDM) Project Cycle. The carbon project will undergo the CDM Project Cycle, which has seven steps: (1) project design; (2) national approval; (3) validation; (4) registration; (5) monitoring; (6) verification; and (7) CERs issuance. These have been described by Mutimba and Kibulo (2013, pp.39–46) as below:

> Project Design: the preparation of the project design document (PDD), using approved emissions baseline and monitoring methodology. It is done by the project participants including the Project developer and Consultant.
>
> National Approval: submitting the project documents to the host country's approval entity-Designated National Authority (DNA) for approval; to confirm whether the CDM project meets its sustainable development criteria; Environmental; Socio-economic; and Technological transfer. The DNA issues the project with a letter of approval (LoA).
>
> Validation: independent evaluation/auditing of a project activity against the requirements of the CDM modalities and procedures on the basis of the PDD. It is done by an accredited designated operational entity (DOE), private third-party certifier. It generates a validation report for submission to the CDM-Executive Board.
>
> Registration: formal acceptance by the Executive Board of a validated project as a CDM project activity. A valid project submitted by DOE to the

CDM-Executive Board with a request for registration; It is also a check for completeness by the secretariat and vetted by the Executive Board.

Monitoring: of actual emissions by project participants according to approved methodology and monitoring plan, continuously during the implementation.

Verification: an independent review of the monitored reductions in emissions by sources of greenhouse gases that have occurred as a result of a registered CDM project activity during the verification period; DOE verifies that emission reductions took place, and the amount claimed, according to approved monitoring plan.

Certified Emission Reductions (CERs) Issuance: DOE submits verification report to CDM Executive Board with request for issuance of the Certified Emission Reductions (CERs).

The project's verifiable baseline is that currently over 90 per cent of energy needs in Kilimanjaro are wood-based, with minimal effort at improving efficiency or switching to alternatives (GEF, 2010, p.33). The success indicator is carbon mitigated from institutions' energy switch and improved efficiencies. The target is at least half a million tonnes of carbon-dioxide mitigated at the project's mid-term; and a million cumulative tonnes at the project's end – within four years from 2010 to 2014 (GEF, 2010, p.33). Another success indicator is the percentage of cooking energy for institutions from alternative sources. The baseline is 100 per cent wood-based cooking energy. The target is at least 40 per cent of cooking energy in the public institutions coming from alternative, and the rest from highly efficient systems (GEF, 2010, p.33).

A survey of institutions showed that at least 51 per cent were aware of carbon credits and 88 per cent were willing to participate in the carbon finance project (Mutimba, 2013, pp.37–38). This indicates high expectations. Yet, there are great risks. Incentives and potential for the project are vested in carbon markets. Should these crash, or should carbon prices fall below those of alternative land/energy uses, the incentives will disappear. Also, the complexity and longevity of the CDM-project cycle is already casting doubts. This is the final year of the SLM project, which includes the carbon finance project. But, according to the project technical advisor, the development of the carbon finance project is not yet complete, and there are no verified emission reductions to report.

Perceptions of 'carbon'

Villagers identify carbon as '*hewa ukaa*' or '*gesi ukaa*', in *Kiswahili*, which loosely mean 'charcoal air'. This expression is used in national climate change sensitizations, implying that the usage was imparted by such campaigns. The expression sticks because charcoal is the most common form of carbon used widely for energy. Other associations that villagers make include 'dirty air', 'dirty charcoal air', and 'dirty industrial air'. They associate carbon with fire, smoke, burning of wood, burning of forests, burning of vegetation, clearing vegetation, etc. Fewer associate it with 'industrial emissions of dirty air'. Thus, the blame on local forest/land users rather than global industrial emissions is sticking.

However, carbon finance/credits are less known. In Camco's survey of institutions, although half of the respondents had heard about carbon credits, they did not understand the details of how they work (Mutimba, 2013). Among the villagers that I interviewed, even fewer understood. A villager expressed his understanding thus:

> Forests can clean the dirty air from industries. So they can protect us from 'changes in the behaviour of the land' [In Kiswahili, '*mabadiliko ya tabia ya nchi*']. So those who plant trees to clean this dirty air get compensated.

Surprisingly, villagers who knew about carbon finance/credits said they heard about it happening elsewhere in Tanzania. Commonly, they cited the REDD (in Kiswahili, MKUHUMI) pilot sites. None mentioned the carbon finance project in Kilimanjaro. One villager, expresesed a concern shared by many: 'Why are the people of Kilimanjaro being forced to keep their forests to clean the dirty air, without compensation, if there are such compensations?' (Mosha, 2012).[6]

The interviewee mentioned being aware of such a project (MKUHUMI/ REDD) in the Shinyanga region:

> Shinyanga is very dry. It is like a desert. They hardly have trees there. But they are being compensated. Here, we have kept trees on our land since the time of our grandfathers and fathers. Why are we not getting compensated? Why are we being forced to keep trees on our lands, and not benefit from them? I have planted 300 trees which are now old. But I cannot cut them. I am also old, and may soon die. Who is going to benefit from the trees I planted?
>
> *ibid.*

This concern has been repeated in different forms in all nine visited villages in three districts (Moshi Rural, Hai and Rombo). The carbon finance project has hardly made contact with ordinary villagers. But fortress carbon management is felt everywhere.

Diverse livelihood interests and project impacts

Beyond carbon, SLM project officials reported that they are also undertaking other livelihood activities, including development of efficient irrigation technologies, and generating income via micro-enterprises. The latter includes beekeeping; negotiating with coffee farmer groups to produce a special kind of coffee that fetches a higher price; and a mushroom project. But these few projects are spread too far and wide to to be noticed – let alone benefit – a significant proportion of rural dwellers. In all the villages I visited, such livelihood projects were unknown.

The carbon finance project is based on the energy switch principle, hence it works with institutions most likely to achieve this shift (Mutimba, 2013). In November 2013, project workshop participants in Moshi voiced two concerns: first, too few public schools were involved, and second, local communities were excluded. Subsequently, additional institutions were added to the project, increasing the total population involved. But this is still only ~1.5 per cent of the population size of the Kilimanjaro region. At best, the project will serve as a demonstration of energy switching, and will not have a major impact on the overall situation.

Regarding villagers' exclusion, project officials felt that villagers are unlikely to be able to meet the project's rigorous 'technical requirements', notably record keeping. Also, the high transaction costs of bundling small individual emission reductions to achieve large sellable bundles may likely be prohibitive. The selected institutions will act as pioneer innovators, and inspire others to copy, or inform future policy, they argued. A suggestion was made to involve villagers through their cooperative societies, as 'community-based-institutions', but this has yet to be adopted. Thus, as a direct result of the project design requirement to gain carbon credits, all local level stakeholders have so far been excluded, except as represented through formal state institutions. The argument was that involving other potential beneficiaries was too costly and the systems for carbon emission reduction validation too complex. The prospects of the proposed later scaling up look dim.

In addition to not being involved in the technology-led energy switch programme, villagers are being coerced into changing their forest use practices, including for energy provisioning through fuelwood gathering. Due to new regulations, pushed as a result of the project, they are prohibited from using even their own (community or individual) forests and trees. A project official believes that this will work by encouraging those who previously used public forests to plant trees, which will benefit them when the ban is lifted, and by encouraging the switch from wood energy to sustainable alternatives (SLM-Kilimanjaro Technical Advisor, 2014).[7] Yet, contrary to these expectations, a recent study concludes:

> With more than 70% of the people engaged in small scale agriculture, household incomes are low with 85% of the people earning less than Tsh 300,000 [~190 USD] per month. Poverty is particularly high in the rural areas where 87% of the people live. There is a link between income and the choice people make with regard to the choice of fuel […] not surprising that most of the people here use firewood and charcoal because these fuels are affordable and available […] Despite the many programs that have over the years attempted, to promote dissemination of improved cooking technologies, adoption rate is low at 3% of rural households.

Mutimba, 2013, p.51.

The strains of fortress carbon management are also felt by institutions:

> The existing ban on tree cutting on private land and restrictions to access firewood from government reserves has created a shortage that has led to increased firewood prices. Institutions are struggling to cope with the situation and have had to divert funds earmarked for other expenses to the purchase of firewood. The situation is unsustainable and might soon implode with serious consequences including closure of some institutions.
>
> *Mutimba, 2013, p.52.*

The strict rules of the 'fortress conservation' approach are causing real hardships, and multiple challenges. This is having implications for resource tenure and access.

Impacts on resource access

> In my entire life I have never seen this use of force to protect a forest [...] Even the colonial government was better, because it gave us the half-mile-forest for our subsistence uses.
>
> *Mallya, 2012.*[8]

This lament of a villager who resides near the alienated forest signifies the local struggle. Villagers and their leaders view the alienation as undemocratic, and outcomes as unfair; both lacking responsiveness to their needs and accountability. Prior to the alienation, there were neither prior deliberations by village councils and assemblies to endorse the decisions, nor consultation, nor prior free and informed consent, nor announcement of the decision to villagers. Villagers became aware only when they were evicted from or harassed in the forest by park rangers. The results of a social survey (Table 3.1) show that the alienation has had major livelihood consequences for local communities due to the restrictions imposed.

Previous forest authorities and district councils had introduced – albeit limited – downward accountability to villagers via joint management agreements. But, being a parastatal organization, the Kilimanjaro park authority's codified accountability is upward to the central government via the Tanzania National Parks Authority. This lack of downward accountability to villagers is further enforced by state bureaucrats, such as the Regional Commissioner, District Commissioner and district and village executive officers.

Villagers' defiance of impositions has been met with brutal hostility and coercion by park rangers. Often the rangers are alleged to ask for bribes from villagers found in the forest. Incidences of physical and sexual violence on villagers in the forest have also been reported (Kijazi, forthcoming 2014). In a social survey, violence and coercion topped villagers' problems of forest access and use (Table 3.2).

TABLE 3.1 A survey of impacts of forestry regulations on villagers' priority forest needs

n = 60. Ranking aggregation by Borda Count Rule.

Category and Rank	Prior to Half-Mile-Forest Alienation	Current Status
I. Fuelwood	Free access in the	Women only, no use of
II. Livestock Fodder	half-mile-forest strip	tools (axes and bush-knives)
III. Water	Regulated by village leadership and customary leaders	Maintenance of irrigation canals require permit from Park Authority
IV. Building Materials	Regulated access by village leaders	Prohibited
V. Gathering – medicines, seeds, nursery soil, etc.	Regulated by village government	Prohibited
VI. Beekeeping	Regulated by village government and district natural resource office	Prohibited
VII. Cultivation/ taungya	Not allowed in half-mile-forest – allowed in industrial plantation zone in Rombo district	
VIII. Lumbering	Allowed under district permits	Not allowed in half-mile-forest – allowed in industrial plantation zone in Rombo district
IX. Hunting	Prohibited	

TABLE 3.2 A survey of villagers' problems with forest resource management, access and use

n = 60 (30 Males, 30 Females). Ranking aggregation by Borda Count Rule

Category	Overall ranking	By males	By females
Coercion; violence by rangers	1	1	1
Restricted access to the forest (men completely lack access)	2	2	3
The ban on the use of tools in the forest (axes and bush-knives/pangas) by women	3	4	2
The tree-cutting ban on private farms	4	3	4
Corruption and forest degradation under park authority	5	5	5

Responding to villagers' livelihood and coercion based concerns during an interview, the RC stated:

> When someone is sick what do you do if you are a doctor? The solution may be twenty injections. The patient may not like twenty injections. But you have to give them twenty injections. They will first complain. But when they heal, they will thank you. Kilimanjaro is sick, the environment has been destroyed. By taking these actions, we are also helping the people of Kilimanjaro.
>
> *Kilimanjaro Regional Commissioner, 2013.*[9]

But the villagers are not just lying down and taking these impositions. Resistance of different sorts has been growing.

Resistance, subversion or accommodation?

The state's coercive apparatus attempts to enforce fortress conservation in the name of carbon capture is countered by villagers' resistance. For example, two blockades were staged at Mweka and Marangu tourist routes in 2009 and 2012 respectively. Both were protesting about the violent treatment of villagers by park rangers. The government reacted by sending the armed Field Force Unit (FFU). In Mweka, villagers brought their children and dared the soldiers to 'kill them with their children'. Mediation by a district official resolved the tension. In Marangu, the FFU fired tear gas, and villagers responded by throwing rocks while hiding in banana fields (Kijazi, forthcoming 2014).

Disenchanted villagers have also resisted elite capture. In the Moshi Rural district there were two incidents of timber confiscation by villagers. They blocked the road to stop the suspected vehicles, and punctured the tyres. The drivers escaped. The villagers called district officials, who auctioned the timber, and the revenues were shared between the district and village governments (District Natural Resource officer, 2013).[10] In one village, the entire elected village environmental committee wrote a letter of resignation to the district executive director. This resignation was intended to protest against the Regional Commissioner's appointment of a parallel environmental committee led by a division officer (appointed official) and a division militia (*mgambo*). The appointed committee came into regular conflicts with the democratically elected committee. The former took the roles of the latter, demanding that the latter stop carrying out their responsibilities despite being elected by local citizens/villagers. These appointed committees and their militia (*mgambo*) feel they are above the well-established environmental by-laws established by the elected village councils. They would often invade villagers' houses and demand to search them for illegal timber, bypassing the established village authorities, and without search warrants. This has created conflict between them and the elected village councils and environmental committees, who feel that they have the authority to protect both the environment and villagers against such harassment.

Such dramatic protests are rarer than the covert daily resistance that is villagers' 'weapon of the weak' (cf. Scott, 1985). This is what one villager described as a 'cat and rat game'. Thus, defiance is widespread. Some women continue to use tools (axes and bush-knives) in the forest despite the existing ban. Some tree-nursery growers have reported declines in seedling uptake, as some villagers now 'don't want to plant trees that they will not benefit from'. I observed overgrown seedlings in nurseries, for example. The ban on men from entering the forest has encouraged arson/sabotage. There are reported cases of destroyed trees in the park by striking nails, or boring holes and putting salt in. Starting forest fires is also suspected. Villagers have also attempted to confront and mock unaccountable officials. During one stakeholder meeting one villager (a retired forester), confronted and mocked park officials:

> During my tenure in the forest service, I protected the forest only with a club. But I was able to get the support of villagers. And we planted trees. Tell me, sir, after years of using the gun-barrel, what do you have to show for it? Shot people? Assaulted women? Answer sir? Can you show us a single tree you have planted in the forest?[11]

During a District Commissioner's meeting with villagers at Kibosho East ward, when asked to discuss the environmental degradation they have seen, the first villager to speak asked the District Commissioner:

> Which forest do you people want us to discuss. The one you stole from us?

Conclusion

Fortress conservation has ended participatory governance in Kilimanjaro. The actors who supported local people's inclusion in forestry included the forestry authority, elected district councils and village councils. They have lost forest resources, revenues and decision-making powers. However, the greatest losers are the villagers. Fortress conservation has excluded them from the forest resource, yet the supposed benefits of carbon finance that has precipitated this change in forest governance has not reached them, as they have been excluded through a different process of project design which finds it too difficult to involve diffuse, poorly educated villagers.

The greatest winner in this story is the park authority, having gained 143 per cent of park area, and near exclusive powers to exclude other actors. The regional officials and other state executives are also winners. As central government representatives, the centralization of forestry in Kilimanjaro has given them more power. The 'green development' agencies (UNDP, GEF, etc.) find themselves in a quandary. In theory, fortress conservation would protect their environmental interests, but in practice, elite capture renders quite the opposite results. Social resistance to these 'development' projects is already derailing the potential gains promised by their green development agenda, including the much hailed carbon finance project.

What we witness here is the 'environment-conservation complex' of Tanzania's political economy, characterized by continuously expanding its protected area estate by coercion; local exclusion from conservation benefits; tourist and hunting revenues captured by the state; and donor interest to fund the success of these estates (cf. Brockington *et al.*, 2008). The climate and carbon sensationalism has simply added new tools to the old fortress conservation toolkit. The park authority, entrusted with the forests of Mt. Kilimanjaro, is a parastatal organization upwardly accountable to the central government. Similarly, Kilimanjaro regional administration office, which initiated fortress conservation on private lands, is a deconcentrated authority of the central government. The collective outcome is forest re-centralization, incited by climate and carbon finance. Carbon finance drives the green developmentalists' agenda, and so it follows that their projects are necessarily centralized and exclusionary, simply because the processes for accessing this money is too complex to involve the wider population. It is therefore not surprising that the carbon project has progressed slowly, and remains centred on a few institutions that are able to take on the requirements of monitoring, auditing and reporting.

This recentralization of forests has led to a significant decrease of the ability of elected leaders to guarantee resource access and democratic representation of their people in forestry. It also has reduced local peoples' abilities to sanction forestry decisions and authorities. It has favoured centralized rent-capture, technical expertise and bureaucratic/autocratic control over democratic deliberation and accountability. Decision powers have been entrusted with park bureaucrats, regional administrative offices, local executives and experts and local elites, diminishing the potential of local democratic practice in resource management.

This chapter, therefore, does not simply add to calls for a more socially just conservation, rather it suggests that carbon-forest enthusiasts and other conservationists ought to consider the often ignored issue of local representation in conservation – as a matter of justice, rural emancipation, as well as a practical basis for achieving their own goals.

Notes

1   The official, who was interviewed at the UNDP/COMPACT project office in Moshi, in March 2013 admitted that the COMPACT project as representative of UNDP made a recommendation to annex the forests into the park, following their 'threats study'.
2   Rahima Njaidi, interview at the Tanzania Network of Community Based Forestry Organizations (MJUMITA), Dar es Salaam, March 2014.
3   Interview with Leonidas Gama, Kilimanjaro Regional Commissioner, in Moshi town, at the Kilimanjaro Regional Administration Offices, June 2013.
4   Interview with Leonidas Gama, Kilimanjaro Regional Commissioner, in Moshi town, at the Kilimanjaro Regional Administration Offices, June 2013.

5 Interview with an anonymous project official of the Sustainable Land Management Project, Moshi, Kilimanjaro, February 2014.
6 Interview with Calist Mosha, member of the Village Environmental Committee, Foo Village, Kilimanjaro, February 2014.
7 Interview with the technical advisor of Kilimanjaro Sustainable Land Management project, February, 2014.
8 Interview with J. Mallya, a 90-year-old villager in Mweka village, Moshi Rural District, October 2012.
9 Interview with Leonidas Gama, Kilimanjaro Regional Commissioner, in Moshi town, at the Kilimanjaro Regional Administration Offices, June 2013.
10 Interview with Moshi Rural District Natural Resource officer, in Moshi Rural District Natural Resource Office, January, 2013.
11 Interview with J. Mrema, retired forester, at Kibosho East, Moshi Rural District, November 2013.

4

CARBON IN AFRICA'S AGRICULTURAL LANDSCAPES

A Kenyan case

Joanes Atela

Introduction

Carbon management projects focused on agricultural landscapes aim to replenish the carbon stocks in farmers' fields. This is claimed to achieve 'triple win' benefits, including increased agricultural yields, climate change adaptation and production of tradeable carbon for additional income. Much literature now emphasizes the potential of such schemes, variously framed as climate smart agriculture, sustainable agriculture and green agriculture, especially for vulnerable rain-fed farmers (e.g. IPCC, 2007; CCAFS, 2009; Zomer *et al.*, 2009; Lasco *et al.*, 2010; van Noordwijk *et al.*, 2010; World Bank, 2010b; Minang *et al.*, 2011; Mbow *et al.*, 2014). However, in practice, achieving such triple wins depends on the local context where projects are implemented, including socially differentiated farmers, their assets, socio-political histories, state policies and previous experiences with state and donor interventions. This chapter analyses how this complex setting affects project perceptions, operations and outcomes in the case of the Kenya Agricultural Carbon Project (KACP).

The KACP is one of the first projects initiated by the World Bank to showcase climate smart agriculture in Africa. Supported by the Bio Carbon Fund (BioCF) (World Bank, 2008; 2010b), the project has worked with groups of smallholders in western Kenya since 2008. This chapter draws on field-level interviews, document reviews and policy process analysis (Keeley and Scoones, 2003) to explore the project's narratives, their relationships with farmers' perceptions, institutions and past experiences, and the effects of project-local interactions on the achievement or otherwise of the triple wins. It finds that, in practice, farmers perceive the project largely in terms of maize production, and focus on issues of access to and control over land and water, paying little or no attention to carbon issues. Meanwhile, state actors focus more on cash crop farming for economic growth than smallholders and associated conservation agriculture for carbon. In this context, achievement of the claimed triple wins is negligible.

The case study therefore highlights that carbon management in agricultural landscapes will only prove successful if the commonly deployed top-down design and implementation approaches are reversed. Farmers need to be fully involved, through inclusive and appropriate information about the links between sustainable farming practices and carbon; clarification of carbon rights; and attention to wider development issues such as water access and secure land tenure for women. Only in this way can carbon management realistically expand opportunities and enhance justice for smallholder farmers.

Case study context

The KACP project sites are in Siaya County, around 30km from the shores of Lake Victoria at 1,400m and 1,140m above sea level. The County is inhabited by the Luo ethnic group – the fourth largest in Kenya (Kenya National Bureau of Statistics, 2009), where social life revolves around predominantly male-headed households. Within a household, the eldest son is entrusted to take on the family leadership and resources to extend the family lineage. Sons have the exclusive rights to inherit land from their fathers. Luo men can marry one or more wives and construct houses for them in a compound referred to as *dala*. A group of homesteads originating from a particular lineage forms a social unit called an *Anyuola* (clan), while a group of *Anyuola* forms a *Gweng* (village) in which elderly men take responsibility for reinforcing community norms and rules. Formal governance units such as sub-locations and locations headed by Chiefs or assistant Chiefs provide a link from village institutions to the central government.

The County has a relatively high population density (333 persons/sq. km) with relatively small land holdings averaging 1.03 hectares. The land is largely under family/customary tenure and is almost entirely (98 per cent) used to grow food crops – mainly maize, beans, sweet potatoes and millet – in a rain-fed system. This rain-fed farming is the main livelihood activity, supplying food, income and a source of employment for about 60 per cent of the residents who provide farm-based labour to their neighbours (Republic of Kenya, 2008). Rainfall variability is a major problem, however; oscillating between 100–2,000mm per annum; this affects crop growing periods (Atela *et al.*, 2014a) and causes periodic difficulties with water scarcity and quality. A large proportion of the population depends either on the wetlands and runoffs from the River Yala that drains into Lake Victoria or on dams constructed by non-governmental organizations (NGOs) (Nyaoro, 2001; Kenya National Bureau of Statistics, 2007). For those who live far from the river, accessing water is a challenge. During dry spells, people – mainly women – walk long distances, sometimes up to 4 km, in search of potable water. The challenges of rain-fed farming have been a major focus for international and national agencies and NGOs, including World Agroforestry Centre (ICRAF), Kenya Agricultural Research Institute (KARI), CARE-Kenya and the Swedish

International Development Agency (SIDA), who have worked with the communities in applied research and practical activities to address soil fertility, environmental conservation and poverty alleviation.

The KACP

This social, agricultural and intervention context provides the setting for the KACP, one of the first World Bank-supported demonstrations of carbon management for African smallholders (World Bank, 2005). SIDA funds 38 per cent of the project costs while the implementing agency – the Swedish Cooperative Centre – Vi Agroforestry Program (SCC-ViA) – contributes 32 per cent of the costs. As part of the farmers' contribution, about 30 per cent of the eventual carbon revenue will be used to fund the remaining costs. The KACP is framed as a triple win solution to the climate, agricultural and livelihood problems facing the Siaya people. The project narrative, as contained in design and justification documents, argues that, in the absence of the project, land in Siaya County (and wider western Kenya) will continue to suffer pressure from agriculture resulting in degradation and poor crop yields. SCC-ViA claims that its work with farmers in this area for more than two decades identified poor farming practices such as slash and burn to cause severe land deterioration, harming food security in the area (KACP, 2010):

> The most likely scenario in the absence of the project will be driven by two dependent issues: The traditional fallowing or shifting cultivation system which is heavily based on nutrient and carbon cycling has declined and will further decline due to rapid depletion of per capita arable land [...] As a consequence of this, the soils will further degrade leading to stagnating yield production and critical food insecurities respectively.
>
> *KACP Project Design Document, 2008, p.10.*

SCC-ViA therefore aimed to build on its long-term agroforestry related work in the area to train around 60,000 smallholders, owning 45,000 hectares of land, in the practices of the new carbon project. These include agroforestry, residue management, cover cropping, tillage and manure management. These practices are expected to replenish soil organic carbon (SOC) as part of delivering the triple wins. The World Bank website communicates the triple win narrative to global audiences:

> Triple Win of Climate-Smart Agriculture put into Practice: Africa's first agricultural soil carbon project changes Kenyan lives: Tom Odhiambo and Maurice Kwadha are small-scale farmers in western Kenya, and they understand all too well the impact of climate change on their local environment and food production. On just one acre of land inherited from his father, Tom and his wife Mary practice improved agricultural practices,

which have enabled them to increase yields and make more money than other smallholder farmers in this hot and dry environment.

World Bank News and Events, 8 March 2011.[1]

Peculiar to this narrative is that it links poor crop productivity in Siaya to poor agricultural practices, but with little linkage to climatic changes. This contradicts the farmers' own explanations that blame declining agricultural yields on increasing rainfall variability emanating from global climate change:

> Many people planted maize in these fields but it did not do well. If you came earlier, you would have seen how short the plants looked because of poor rains. When the rains come in good time, the crops do well and I am able to get enough to feed my family for a longer time. Also when the rain is there, many people plant two times in a year first in February and then in August after the first harvest. However, now we only plant when rains fall but you can't know when.
>
> *Community member, Wagai, August 2011.*

Farmer views therefore conflict with the externally driven project narrative. While the project implementer has worked with the Siaya farmers for many years, and could realistically inform the project design, the project narrative reflects the interests of global actors such as the World Bank in constructing agriculture as part of formal climate change mitigation programmes, such as Reduced Emissions from Deforestation and Degradation (REDD). The World Bank emphasizes that African agriculture must undergo a significant transformation in order to meet the related challenges of food security and climate change (World Bank, 2008, 2011). The consultants it appointed, who were technically responsible for the project design and monitoring procedures, were likely expected to conform to their employer's narrative. Farmers were not adequately consulted in project design, and consequently their understanding of the KACP remained narrow even after six years of implementation. These farmers are primarily interested in boosting production and achieving food security – 'the first win' – and so for them, the project as expected provides a solution to low yields of food crops, especially maize and beans. Indeed a strong farmer narrative revolves around good farming practices for increased maize yields:

> I have been following up the project through community resource persons. I want to engage with the project because it provides skills for new land use practices and be able to get the 5 bags of maize I used to get from my land when I moved to this area years ago.
>
> *Male farmer, Wagai, August 2011.*

Interestingly, while farmers have a strong narrative associating the KACP with good farming practices for increased maize yields, most associate declining maize yields with variable rainfall. Any project that highlights either fully or partly the need to improve maize productivity, however obliquely, is considered relevant. Farmers seem to have interpreted project communications around climate change to suggest that sustainable land management practices in their farms will solve rainfall problems – downplaying causes of climate change in global processes of energy and land use.

The emphasis on maize is a new incarnation of a historically embedded narrative. With maize being the dominant crop, farmers have long been exposed to technologies to improve production, whether new seeds or improved fertilizers. Through numerous interventions from government, aid projects and NGOs, farmers in this area have been made to understand that a 'good farmer' is one who implements 'good' farming practices (i.e. those introduced by external interventions) and gets high crop yields with stores always full of maize. In recent years, a range of prior projects have promoted a variety of additional practices for improving production. This part of western Kenya has been a major soil fertility testing ground for the ICRAF, KARI and a host of NGOs, all of whom have promoted improved fallows, biomass transfer and alley cropping disseminated through contact farmer and demonstration approaches. During public gatherings, such as Chief's Barazas, 'good' (i.e. high maize producing) farmers are conferred prestige by being highlighted as examples.

Similarly, Kenya's Agricultural Sector Development Strategy (ASDS) for 2010–2020 (Republic of Kenya, 2010a) focuses on enhancing economic development via agriculture. Through its focus to promote a quick fix to solve the food crisis of the last five years (see Republic of Kenya, 2010b), the strategy emphasizes agricultural intensification, fertilizer use and agricultural mechanization – all classified as carbon emitters according to the KACP monitoring protocol.[2] The strategy has no explicit provisions for addressing climate change in agriculture.

In this established set of narratives and practices around agriculture projects, there is little or no consideration of 'climate smart', 'green' or related elements such as the carbon stocks and payments associated with a project such as the KACP. The KACP has thus found it daunting to encourage farmers to appreciate the additional two wins of 'resilience' and 'reducing carbon emissions for income'.

Implementation: Expectations and experiences

The implementation process of the KACP revolves around project developers, project implementers and farmers (Figure 4.1) and involves a set of stages: the project development process, introduction of the project intentions to farmers, training farmers in sustainable land management activities and monitoring and reporting the outcomes.

Project development and farmer engagement are informed by standard World Bank guidelines. In these, carbon projects are expected to develop, disseminate and support agricultural practices that achieve 'triple wins' in a participatory way (World Bank, 2011) but also in line with the global standards set by the UNFCCC

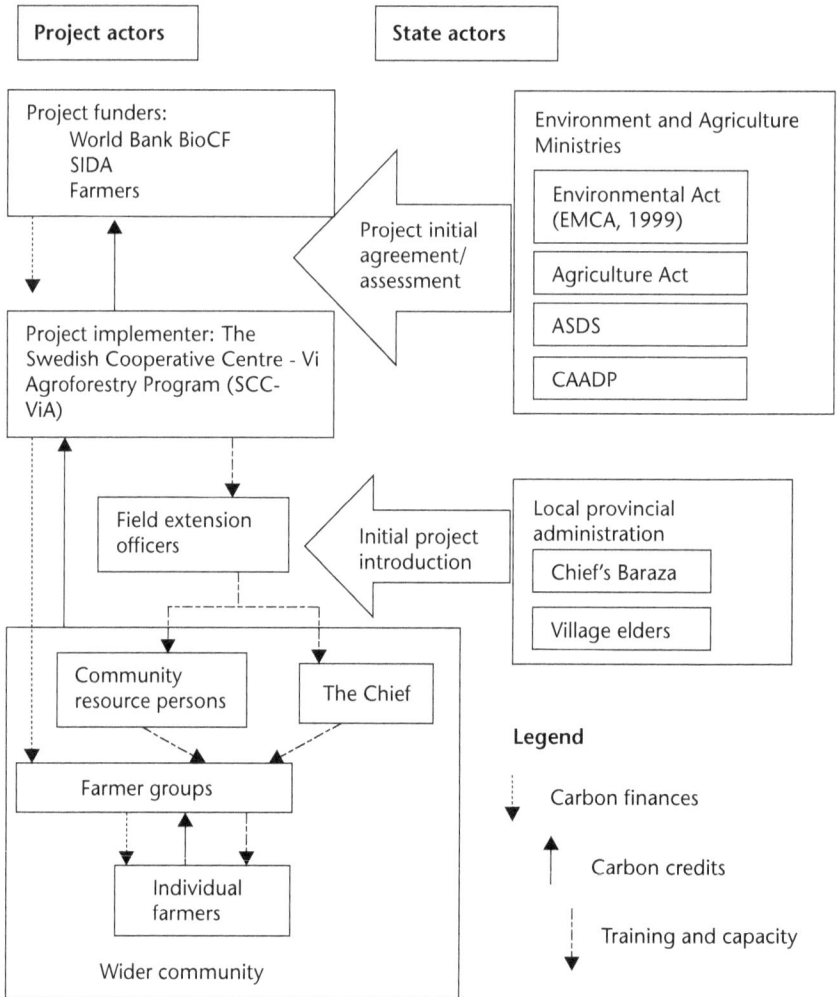

FIGURE 4.1 The organization and implementation process of the KACP

and its land use change and forestry provisions. These expectations formed part of the Emission Purchase Agreement signed for KACP between the SCC-ViA, the Kenyan Government and the World Bank. This partnership agreed on emission reductions by the SCC-ViA and purchase of carbon credits by the World Bank while the government was to provide an enabling environment (World Bank, 2010a).

Nonetheless, at the local level, project implementation and farmers' choices and perceptions of the KACP were significantly shaped by communities' historical engagements with prior projects in the area. For instance, in laying out its intentions, the KACP worked through established contacts and networks around the Chiefs and their Assistants, community resource persons and leaders of various community-based organizations. These contact persons were identified and engaged in a seminar for a few days to familiarize them with the project intentions. Through meetings and teaching sessions organized by these contact people, farmers were told about the project activities and what was expected of them. Most farmers heard about the KACP through meetings organized at the Chief's Barazas, while a few learned the details through other fora, such as umbrella Community Based Organizations (CBOs), community resource persons, farmer visits, trainings and group meetings.

Chiefs' Barazas are a decentralized local administrative forum where governance issues affecting local people are discussed and new development initiatives from the government and NGOs are announced. Barazas were historically used by colonial officers as conduits for exercising external authority over their local clientele (Fleming, 1966). Practices of introducing initiatives to farmers through local elites in fora such as the Chief's Barazas have evolved over time, but have commonly been deployed by external projects as an effective way to gain farmer acceptance. In the post-colonial era, the Barazas have frequently been utilized as conduits for conveying farm inputs such as maize seeds, fertilizers or relief food from the government to the farmers. Consequently, most farmers in Siaya County are familiar with and often attracted to Barazas, either as willing participants or in expectation of new benefits. While Barazas provide an opportunity for projects to 'gain acceptance', farmers tend to perceive them as a top-down form of participation mediated by local elites and contact farmers who may shape the nature and content of externally designed projects, and even coerce involvement. Indeed, as a Chief put it:

> I represent an authority in this area and without my permission no project can be approved by the community. The community has trust in me and believes I understand all aspects of governance and development. Therefore any project coming into this area has to go through my office.
>
> *Chief, West Gem Location, Wagai, August 2011.*

Following these introductions, farmers were then contracted to be part of the project by signing 'farmer commitment forms' as a commitment to implement carbon-focused sustainable land management practices. Extension staff were expected to train and backstop farmers in the implementation of sustainable land management practices including agroforestry, alley cropping and mulching, all believed to be linked to the triple wins. These extension staff mostly drew instructions from the KACP management based in the projects' offices in Kisumu City, which in turn referred to the project consultant to ensure that activities were operating within the desired standards. As one extension officer put it: 'We still do not understand so much about the carbon project. But we work with community resource persons to implement what the officers have developed' (KACP extension staff, Siaya, November, 2013).

While extension staff are deployed within local areas, giving potential for close contact with farmers, in practice their working conditions are difficult. They are few, relative to the number of farmers they are supposed to work with – for instance only three extension workers were stationed in Wagai and were expected to work with 600 farmers each. KACP extension officers gain little assistance from government extension officers stationed in the same area: while the government was part of the KACP deal, and committed to provide enabling conditions for the project, government extension officers were not involved in the project, knew little about it and claimed that it was difficult to visit farmers because the government did not facilitate them with fuel and lunches. KACP extension staff claimed that government staff always want money even to perform simple duties, yet they (the KACP staff) must depend on carbon money that takes a long time to materialize, and is significantly affected by global market forces.

In this mix and laxity in extension work by both the project and state, it is common for community resource persons to assume the role of contact farmers and mediate between extension staff and farmers. Through such advantageous positions, contact farmers usually get first priority in access to project resources and initiate demonstrations on their own farms, which other farmers are expected to observe and replicate. The decision of a farmer to be part of the KACP also relies on him/her being part of a group. However, not all group members embrace project activities. Within groups, farmers were observably more enthusiastic about a 'saving and lending' initiative compared to the sustainable agricultural activities. The perceived economic benefit of engaging in an initiative is often paramount, not because these farmers are business actors but because they need money for pressing food purchases, especially during poor harvests:

> Sometimes you visit a farmer and you find malnourished children desperately looking at their mother and so you wonder whether to start the training on carbon work or sort this social problem first, and in such a circumstance, the farmer normally presents his/her urgent need for food in a way that easily overrides the aim of the visit.
>
> *KACP extension officer, Wagai, August 2011.*

The KACP's carbon accounting and monitoring are guided by an approved Verified Carbon Standard (VCS) methodology that combines the Roth-C model with Activity Baseline Monitoring Survey (ABMS) (Bird, 2012). Farmers participate in the ABMS part of the monitoring, and this is supposed to feed into a periodic validation (2011, 2014 and 2017) process during the 20-year project period. A group of farmers has been selected to be part of the ABMS, and to provide a lens through which other farmers' activities can be monitored. The ABMS farmers' record of the sustainable land management practices they implemented during a particular period, and the outcomes in terms of crop yields and other factors, are submitted to the project management for further technical accounting. This makes the monitoring process reasonably participatory. However, the ABMS farmers were technically selected through scientific procedures of random sampling aided by a computer simulation system and located using GPS coordinates. They were not consulted on their willingness to be part of the process, compromising the principle of free prior informed consent as required for carbon trading projects (Streck, 2012). Some ABMS farmers only found out that they were part of the sample when extension officers and project surveyors appeared on their farms to record land uses. Not surprisingly, some selected plots fell on areas where the farmer was an immigrant who is only a land tenant. Land under short-term lease arrangements is not eligible for carbon-related activities, according to the rules and regulations of the BioCF (KACP, 2008). Further, most farmers do not understand the contents of the 'self-evaluation' form used in monitoring, since they are written in English. Even extension staff regretted their inability to assist farmers adequately to understand the procedure because their input into its design was minimal. In this context, some farmers at the beginning of the project gave false figures of land acreage at their disposal for fear that the project would take away their land. Others understated the yields from their farms in expectation of food aid from the project, while well-informed farmers tended to overestimate yields and activities in expectation of being valued as 'good farmers', eligible for prestige during public gatherings – or more shares of carbon money. Such expectations of external assistance, as already highlighted, are informed by long histories of intervention in the area.

These sampling and transparency issues add to the uncertainty around agricultural carbon finance. Uncertainties are accepted as part of the project design. This is why the project discounts 60 per cent of the accounted carbon. However, this discounting assumes uncertainty in one direction (underestimation) – despite the fact that field experience shows that farmers can overestimate values as well as underestimate them. Moreover, the KACP works in an area where several other NGOs and government organizations exist to support farming and environmental conservation. KARI for example, through the Western Kenya Integrated Ecosystem Management Project (WKIEMP), until 2010 worked with the same farmers to establish community tree nurseries and undertake participatory reforestation. Some of these activities are still ongoing and KARI do not value any carbon resulting from the activities they promote. During monitoring under the KACP, farmers tend to record all activities in their farms, regardless of which project supported them. This means that some carbon is added into the KACP basket by default. Taking away a massive 60 per cent of carbon credit allocation to farmers is therefore not adequately justified. Carbon accounting drives the process of who benefits and who loses, yet it is clearly contested in practice. Such flaws may benefit other actors and networks more than farmers, reflecting the power relations discussed earlier.

The experiences of introducing and engaging farmers in the KACP therefore illustrates that implementing carbon projects in smallholder farms is highly contested, subject to complex negotiations and affected significantly by past interventions in an area. New projects must respond to this context, and so it is no surprise that a carbon intervention mimics prior interventions that have shaped community understandings in particular ways. While such prior interventions provide lessons for the new externally standardized and market-linked carbon projects (Blom *et al.*, 2010), the level to which these efforts promote farmer capacity for informed engagement is questionable. In the KACP area, many farmers have little or no idea that the new project is being implemented under a 'triple win' climate smart agriculture banner, and they assume simple continuity with past interventions. This might be a short-term response, and, over the long-term (20 year) KACP implementation period, it might be expected that local understanding, representation and empowerment of communities to expand their options in the face of climate change will increase. But whether or not this is the case, farmers have the right to know the context for carbon interventions and to share information about potential carbon revenues being developed through practices on their land. Otherwise, political and ethical conflicts over disclosure, accountability and transparency will inevitably emerge.

Meanwhile, with the project resulting in shifts in rights (over land, trees and carbon), social differences emerged to complicate these concerns further.

Gender and generational issues in the KACP

Farmers' participation in the project shapes and is shaped by social issues related to gender and generations. In the KACP, as we have seen, farmer groups link the contact persons and the individual farmers. Notably, women largely initiate the formation and running of these groups and, in most cases, dominate in terms of numbers. Analysis of nine groups (Table 4.1) reveals that, on average, women constitute about 75 per cent of the groups' membership, with men and youth sharing the remaining 25 per cent. This disparity largely reflects gender roles, given that women in this area are the primary food producers. As one put it: 'I felt that coming together provides a collective power to face the daily challenges that we face in this community' (female farmer and a member of local women's group, Wagai, August, 2011).

While women are the farmers in these areas and take the lead in the communal groups, most resources – especially land – are traditionally controlled by men. Men have greater authority in deciding which farming methods and agroforestry systems to use, and even which crops to plant. Although women dominate the groups through which carbon related land uses are disseminated, it is nevertheless men who are expected to approve their implementation. This reduces the extent to which project activities disseminated through group training or demonstrations take place in practice.

Young people's participation in agriculture, let alone projects related to carbon, is as low as 5 per cent. The poor involvement of young people in group-based rural farming is partly conditioned by cultural norms around age sets, which make it difficult for young people to associate with elderly women and men in the groups. Youths are also reportedly impatient and prefer quick money, which in most cases

TABLE 4.1 Membership composition of the groups working with the KACP

Group name	Men	Women	Youths	Total	Able to read and write	Able to explain triple wins
1. Kobeto	3	26	1	30	7	2
2. Ahono	6	22	28	56	20	6
3. Kanyanga	4	22	0	26	4	2
4. Tich tek Te	0	29	1	30	0	4
5. Kamsao	0	23	7	30	10	6
6. Ogwedhi	1	25	2	28	27	0
7. Wacho gi timo	1	23	7	31	1	3
8. Kinda	4	21	5	30	15	3
9. Liech Gumbo	5	23	2	30	5	1
% total group members	8.2	73.5	18.2	100.0	31.2	9.5

is not forthcoming from environmental initiatives such as carbon projects. Indeed, the majority of young men in the project area prefer operating motorbikes which generates immediate income. According to village elders, there is even a trend for youth to sell local resources such as land and then migrate to urban areas to look for formal employment. This form of age-related gender disparity potentially bears negatively on the KACP in the context of the 20-year implementation period. At the moment, the average life expectancy in Siaya County is 50 years (Kenya National Bureau of Statistics, 2007). Most farmers are elderly women nearing this age, who in 20 years would be 70 – way above the life expectancy level. In reality, most farmers currently in the groups may only be active during the next five years. It is the youth who are expected to continue the good farming practices promoted under the KACP, yet their thin presence in the groups may negatively affect the project's sustainability, given its 20-year implementation context.

These gender and generational issues, alongside other experiences in implementing the KACP, in turn have significant implications for the livelihood and ecological impacts of the KACP – which are decidedly mixed.

Implications for triple wins: Who benefits?

The implications of the KACP can be considered in relation to each of the claimed triple wins: carbon rights and benefits, resilience and agricultural productivity. The project documents (KACP, 2008; Wekesa, 2010) claim that 490,500 tCO2e will be generated after discounting 60 per cent to cover uncertainty during the 20 year project period. The World Bank's BioCF is already purchasing part of the 150,000 tons of emissions reductions that it committed to purchase at approximately US$600,000 between 2009 and 2016.

These expected carbon revenues and their benefit sharing procedures were, however, not clarified to farmers at the start of the project. Even by 2014, farmers remained unaware of how much they could expect as their carbon revenue share.

The project managers explain that they keep carbon payments off the agenda to avoid raising expectations, only highlighting them as 'a bonus' for implementing the good farming practices that the project teaches. In the context of this oblique information on carbon revenue, the project received the first payment for 24,788 tCO2e in late 2013. The actual payments were not disclosed to farmers but several indicated that payments were made to groups. The highest paid groups received 6000 Ksh (US$70) while the lowest paid group received 1900 Ksh (US$24). This was the first payment farmers had received since 2010 when they signed their commitment forms to start sustainable land management practices on their farms. If divided amongst a group comprising 30 people, each member of the highest paid group would receive 200 Ksh (US$3) and of the lowest paid, 62 Ksh (less than a dollar). Project staff explained that this was a year's payment and additional revenue is still expected. Even so, a projected calculation indicate that the farmers' share will significantly be reduced to only 28 per cent of the total carbon revenue due to uncertainty accounting and their contribution towards extension wages. The carbon income is therefore insignificant to the livelihood of farmers who require an average of US$1.2 a day for basic living. Groups have therefore decided to invest their earnings collectively, either in group farming or tree planting, but even so, the impact is small. KACP is one of the first pilots for agricultural carbon management and it may therefore be reasonable to seek to avoid false expectations. However, later contestations may arise if farmers – as the producers of carbon – argue that they had no input into designing the sharing procedures, yet are expected to implement the project's activities and suffer reinforcement measures such as reducing a farmer's shares to compensate for absenteeism during a joint farming session.

Aside from direct earnings, the project implicates issues of carbon and resource rights. The carbon accounting procedure in the KACP is based on individual land holdings, as is the concept of 'carbon rights' and associated payments. However, more than 50 per cent of land in the project area is held customarily and legitimized by traditional passage of use rights from one generation to the next (KACP, 2010). Customary land rights are held by individual families, but communal use of the land is a common practice. Given that residue incorporation and vegetation retention in these farms are some of the carbon-generating activities, should farmers allow free grazing of land during the dry season or instead conserve residues for sequestration, and individual benefit? Such conflicting land and resource tenure arrangements may create significant conflicts as the commoditization of carbon creates incentives to privatize and individualize resources. This potentially aggravates local injustices in terms of access and ownership of land and resources, potentially excluding tenant farmers, landless people, women and youth who traditionally have no ownership rights over land in this part of Kenya, as in many African societies. Appendix 4 of the project design document (KACP, 2008) attempts to address these rights in line with BioCF rules of engagement. In this appendix, the project recognizes customary land rights but ignores the informal yet very important communal land uses common in the area. Claims to carbon rights by farmers owning land under such customary arrangements may not be legitimate in Kenya's court of law. The transfer of land ownership is normally based on oral commitments made by fathers to their sons. This implies that disputes over carbon may only be resolved by local authorities, reinforcing biases and power dynamics at the local level.

Despite the land tenure mix, most farmers appreciate the project's efforts to teach them sustainable land management practices. They highlight this as the main impact of the project, even though they do not link these activities with climate change. Many farmers feel that, through the KACP, they are now relatively committed to sustainable agriculture and that their farms have benefited from multipurpose agroforestry tree species. However, it is not clear if these activities have improved yields of maize, the staple crop so important to food security. Moreover, the growth of the trees is affected severely by drought conditions. Farmers ranked access to water as the greatest challenge towards their adoption of agroforestry practices (Figure 4.2). Water scarcity, alongside the gender and economic issues discussed above, was ranked as the most significant factor affecting sustainable agricultural activities. With increasing water scarcity, crops, planted trees and fallows dry up more frequently and women have to spend more time fetching water for domestic use at the expense of implementing sustainable land-use activities. Addressing such underlying critical and sometimes non-climatic factors is an important strategy to enhance farmers' abilities to cope with climate change, particularly in a resource-constrained smallholder farming context. However, during discussions with farmers and a review of project documents, it was not clear if there are efforts in place by KACP to address this underlying threat.

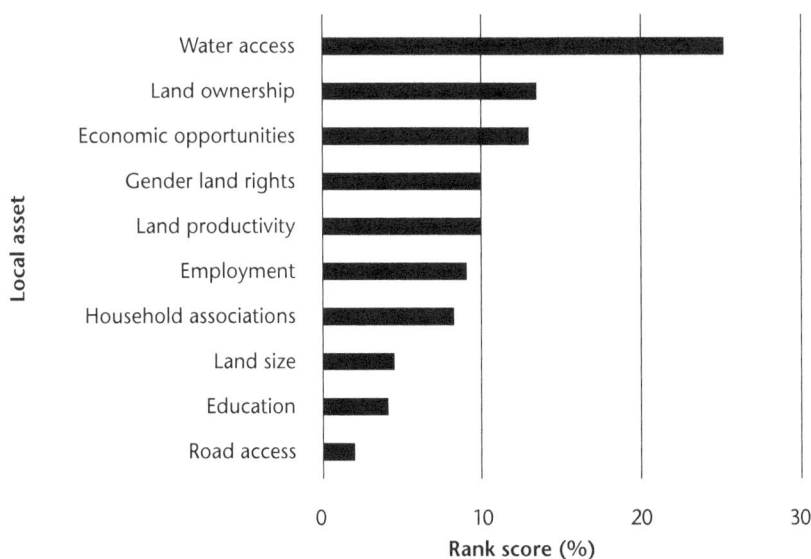

FIGURE 4.2 Ranking of the most important assets for the adoption of project practices

Conclusions

Achieving triple wins through carbon management in agricultural landscapes is possible in theory, but in practice influenced heavily by the complex socio-economic, political and cultural context within which a particular project is implemented. This context includes state policies; prior interventions; farmers' livelihoods and asset holdings; economic and tenure relations; local organizational processes; and gender and generational relations. In the case of the KACP, these factors combined to limit the carbon-related wins, and to restrict even agricultural and livelihood benefits to a few.

In western Kenya, state policies towards agricultural commercialization and mechanization impede state commitments to support triple win agendas in carbon projects, whether logistically or technically. State policies favour subsidies to established cash crops such as tea, coffee or horticulture, to the exclusion of rain-fed peasants who mainly farm for food and household income. This has left farmers more vulnerable to climate change, and with a tendency to view any incoming interventions as a quick fix opportunity for their vulnerable livelihoods. Such perceptions, expectations and beliefs feed into the KACP, and farmers simply expect the project to solve maize yield problems, paying little attention to its carbon emphases of broader information and empowerment aims.

Transformations in approach will be required if projects like the KACP are to make sense to smallholder farmers. These will need to include support to people's access and control over assets such as land and water; redressing gender imbalances in resource access and decision-making, and widening access to alternative livelihood activities. Committing more resources and effort to such broader issues could yield greater benefits than investing in training farmers to implement good farming practices that they have long known through prior initiatives.

Climate science shows that smallholders are not a significant cause but rather a significant victim of climate change. Therefore, social justice calls for compensation, rather than making them bear the cost of mitigation programmes. Commercialization of agricultural carbon finance makes issues of carbon rights a grey area. Capacity building should aim to clarify to farmers the link between certain farming practices and carbon and climate change. Such understanding could help to broaden and enhance farmers' already innovative ways of tackling climate change. Understanding carbon issues will also help farmers to make informed choices and inputs to project activities. Unable to engage with carbon project framings, farmers' knowledge is often sidelined and their inputs to project design, monitoring and benefit sharing are minimal. Democratization of expertise that recognizes farmers' knowledge and input should substitute top-down, science-driven approaches and promote local ownership of carbon management projects in agricultural landscapes.

Notes

1   http://web.worldbank.org/WBSITE/EXTERNAL/TOPICS/EXTSDNET/0,,
    contentMDK:22842518~menuPK:64885113~pagePK:7278667~piPK:64911824
    ~theSitePK:5929282,00.html, accessed 7 October 2014.
2   www.v-c-s.org/sites/v-c-s.org/files/VM0017%20SALM%20Methodolgy%20v1.0.pdf,
    accessed 7 October 2014.

5

'ZONES OF AWKWARD ENGAGEMENT' IN UGANDAN CARBON FORESTRY

Adrian Nel

Introduction

This chapter draws on reflections from two empirical case studies in western Uganda to explore the varied kinds of engagements, exclusions, accommodations and conflicts that arise in the historically constituted landscapes in which carbon forestry projects emerge. The two projects are implemented by large-scale timber plantations companies, which lease land on designated Central Forest Reserves (CFRs) to plant exotic pine and eucalyptus species in the Kiboga and Mubende districts, part of the 'cattle corridor' that passes through western Uganda. Both the implementers, the New Forests Company (NFC – a UK-based company) and Global Woods (a German company operating through their local subsidiary, Sustainable Use of Biomass (SUB)).

The NFC initiated a carbon forestry Afforestation/Reforestation Clean Development Mechanism (A/R CDM) project on its leased land in the Namwasa CFR, which is linked with its holding at the Luwunga CFR. It inherited all the problems of encroachment on the reserve with its lease, and the project implementation has seen the evictions of large numbers of residents amidst a global outcry. In contrast, the Global Woods project on land leased in the Kikonda CFR was at odds with the surrounding community, especially cattle keepers, for ten years before realizing that its strong-armed approach was unsustainable. The carbon forestry project aligned with a more subtle approach and improved relationships and communication with communities. However, the company is complicit with a blurring of the lines between the legal and illegal, in attempting to facilitate its plantation development.

Thus, within Uganda, and across these two cases, carbon forestry project implementation can be highly varied, and unevenly contested. As this chapter shows, local agency and resistance through encroachment are not marginal, but central to the experience of carbon forestry in Uganda. Understanding the different ways in which these resistances are combatted or accommodated – in what Tsing (2005, p.11) calls 'the zones of awkward engagement' between the projects, state bodies and communities – is key to assessing the outcomes of carbon forestry projects. On the one hand, carbon forestry could be said to be complicit with 'expulsion', a concept that moves beyond Sibley's (1995) conception of 'geographies of exclusion' to categorize how negative processes affecting communities (of eviction, dispossession and marginalization) become normalized in carbon forestry's remaking of the 'national' within the 'global' (Sassen, 2013). However, whether this happens in practice is mediated by local geographies and power relations through a process of engagement. While there can be violent evictions in some cases, in others they can be resisted, or tensions between actors can be negotiated and accommodated, as project outcomes are determined in their respective historically constituted landscapes, beyond the script of top-down managerialist policy prescriptions and land-use plans. In these kinds of zones, differences are by no means reconciled, but rather are facilitated or accommodated in fractures and disjunctures, in awkward, non-uniform ways, which, this chapter argues, deserve attention.

I will begin by setting out the contemporary forestry governance context in Uganda, and the power relationship and changing governance contexts upon which the projects are contingent. I will then proceed to explore the projects, and their relative costs and benefits for local residents in more detail, before turning to the ways that project outcomes are mediated in zones of awkward engagement. I will use Biddulph's (2011) conception of an 'agency dilemma' as a starting point. This concerns whether actors can work behind the 'policy façades' to navigate tensions in project implementation, and effect what they see as necessary changes on the ground.[1] Finally, in concluding the chapter, I will interpret the findings in terms of their implications for the three potential ways of interpreting carbon forestry outlined in Chapter 1. I will argue that there is both scope for interpreting carbon forestry as a 'green grab', but also a need to embrace the agonistic politics of dispute and deliberation around resource use and rural livelihoods.

The carbon forestry context in Uganda

Ugandan forestry landscapes are shaped by institutions, political economies and struggles over resources at various scales (cf. Ribot and Peluso, 2003; Robbins, 2012). The emerging forestry sector has evolved from an extractive colonial forestry regime, through periods of disruption and faltering decentralization, to the neo-liberalized forestry dispensation that pertains today. In the sector, as in others in the country, the use of foreign investment for development has been institutionalized and championed, at both the national and international level, and in both policies and processes (Mamdani, 1987). Plantation forestry and private sector investment has been identified as key to protecting natural forests, within the National Forestry Policy (2001) and the National Forestry and Tree Planting Act (2003), both conforming to the countries' Poverty Eradication Action Plan (PEAP) of 2005.

As elsewhere, different processes of upscaling, downscaling and outscaling (involving processes of privatization, deregulation and decentralization) of governance capacities and scales of activity (Vaccaro et al., 2013) have come to characterize changes in forestry governance. An 'upscaling' of authority and agenda setting in policy has occurred due to donors and transnational bodies such as the Forest Stewardship Certification (FSC) body, the United Nations Framework Convention on Climate Change (UNFCCC) and the World Bank's Forest Carbon Partnership Facility (FCPF). Regarding outscaling, the reform of the old Uganda Forest Department under the donor-funded Forest Resources Management and Conservation Programme (FRMCP, 1999–2003) project, has created a new set of institutions to privilege private planting. The National Forestry Authority (NFA) is tasked with commercial tree planting on Central Forest Reserves, the underfunded Forest Sector Support Department (FSSD) is meant to provide oversight for the Ministry of Water and Environment (MWE) and the bulk of forest resources are relegated to the control of a quasi decentralized and largely dysfunctional District Forest Service (DFS). This has been accompanied by processes of downscaling, whereby the responsibility for carbon sequestration is outsourced to corporate actors running industrial plantations and NGOs and carbon providers who broker and sell carbon credits in the Voluntary Carbon Market (VCM). The resulting local-level carbon forestry projects are widely differentiated, from large monoculture reforestation initiatives such as New Forests' and Global Woods', to Ecotrust's Trees for Global Benefit project that deals with indigenous tree planting on 'private forest lands'.

The overall trend in Ugandan forestry has been the shifting emphasis, policy focus, and allocation of resources from territorial forms of governance (with an emphasis on individual forest territories and the maintenance of the 'forest estate') to flow-based governance, administered through the control of flows of commodified biomass (of monoculture plantations and carbon) and the populations of trees and people that interact with it, under 'neo-liberal' forms of governance (cf. Sikor et al., 2013). The policy focus is now on public–private partnerships in timber planting, and on carbon forestry, while budgetary support for local administration costs of supporting central forest reserves has been declining. As Igoe (2010) points out, designated 'protected areas' are less than protected if they house economically important fossil fuels and minerals. The utilization of areas of the Murchinson/Kabalega National Park, and a Ramsar wetlands site in Bulisa for oil exploration; the degazetting of the Kalangala Islands CFRs for biofuels planting; the occupation of Namanve Forest Reserve and others by politically connected individuals (Anti–Corruption Coalition Uganda 2010); and the repeated attempted degazetting of Mabira CFR for the Mehta sugarcane corporation, are all examples of uneven protection and utilization. Similarly, after the process of forestry reform, and primarily due to capacity shortages, CFRs outside of those cherry-picked for the private planting and public–private partnerships (with the support of the Small Production Grant Scheme (SPGS)),[2] as well as forests outside of the protected forest estate, have been marginalized in policy and resource provision.

These changes unevenly reassert and in other instances weaken the integrity of forest territories and resources in the country, and similarly re-emphasize the ongoing exclusionary relationship between protected areas ('fortress conservation') and 'communities'. The line between coercion (involving evictions of so called encroachers) and governance (where tensions are effectively dealt with) is well trodden in Ugandan forestry, and the Uganda Wildlife Authority (UWA) and NFA approach the issue of encroachment on central forest reserves with some trepidation, acknowledging the problem, but at the same time attempting to cope with it as an everyday reality. Indeed it is the territoriality of the 'forest estate' in Uganda itself that is particularly contested (Nel, forthcoming), as CFR boundaries are porous, and improperly demarcated or policed. At the same time, alternative claims to territory subtly or overtly act as a challenge to this territoriality, and of state itself, presenting alternative land-use options and ecological knowledges (*idem*).

The pressures for encroachment include the recent history of violence; unreconciled differences and conflicts between ethnicities within and across the national borders; the spectre of increasing rural landlessness and its connection to poverty (Ainembabazi, 2010); and the location of protected areas both on and near areas of high agricultural potential, making them highly susceptible to conversion to agriculture (Nakakaawa *et al.*, 2010). This pressure is coupled with the compromises, concessions and political economy of the state in 'regime survival' mode, where it is not politically expedient to sanction evictions, unless of course there are economic rents for the elite involved (see below). This is all compounded by increasing corruption at all levels within the NFA. However, as with Uganda as a whole, where the problems of violence, patronage and corruption cannot simply be reduced to moral deficiencies, the problems in the sector stem from the social conditions of forestry, where, despite regular reformulations and revisions, it remains unclear whether forestry policies, laws and territories are acceptable to the local people and appropriate to the local situation (Turyahabwe and Banana, 2008). The problems of the maintenance of contested forest territories are exacerbated when funding is shifted away from local governance capacity building to the facilitation of public–private partnerships, and the funding of NGO activities and carbon forestry. As it stands, donors – most notably Norway – have pulled funding from the NFA following a corruption scandal in 2009, and are unlikely to fund the maintenance of CFRs that would include large scale evictions. This is because of donor accountability to their constituencies at home, and the media spectacles such activities attract; just as human rights concerns with regard to evictions are on the public agenda in Uganda itself. Human rights and environmental watchdogs[3] impress the need to balance livelihoods and nature conservation and prohibit the human rights violations that occur during evictions (Mugyenyi *et al.*, 2005). Thus, it is within this complicated forestry context that the two case study projects take place.

Two projects in protected areas in Kiboga district, Western Uganda

Three of the 'big four' industrial plantations companies in Uganda run carbon forestry projects, including the Norwegian Green Resources AS (and its local subsidiaries), Global Woods (Germany) and New Forests (UK). The last of the four, Nile Ply, is both the only local company and the only one without a carbon forestry project. The foreign plantation companies are primarily supported by private equity funds and large institutional investors for their planting of exotic pine and eucalyptus plantations, but supplement what they describe as narrow profit margins with novel, additional carbon finance revenue streams from the World Bank's A/R CDM and voluntary markets, using the justification that they are reforesting what they describe as 'degraded forests'. The London-based New Forests Company leases two Central Forest Reserves in Uganda, and draws funding from a speculative agricultural fund Agri-Vie, HSBC and the International Finance Corporation (IFC) of the World Bank for its carbon forestry project on one of the reserves. The second, Global Woods, is a German company that operates through its local subsidiary, SUB, and draws funding from the VCM for its Kikonda Reforestation project (which it markets either directly online or through brokers).[4] The project is run on the Carbon Fix methodology (acquired by the Gold Standard in September 2012) for the reforestation of degraded landscapes, and is Community, Climate and Biodiversity Alliance (CCBA) accredited. In fitting with the direction of forestry policy described above, the two companies running the projects have received preferential treatment, tax concessions and support for initial plantation development, as well as cheap lease conditions.[5]

These two projects are part of expressly 'for-profit' plantation activities. They are in some ways also simpler than many NGO projects, because their developmental claims are primarily limited to narratives of job creation and community development projects, and they only work on limited, specifically tailored social and environmental impact assessments (for instance, the Kikonda project vastly understated the number of park adjacent residents, and even ignored the consultants' warnings in reports). While the companies' timber units retain essentially the same materiality throughout the commoditization process, the carbon they produce is (re)imagined in very specific ways. It is both equivalent to emissions reductions in distant places, and generative of community and environmental benefits that only make sense in the context of national forestry narratives and claims to territory, and in relation to the baseline 'without project scenarios' that the project proponents generate. The numerous measurement and monitoring, review and verification (MRV) procedures and protocols that legitimate the production and sale of carbon from the projects are undertaken in line with the Carbon Fix and A/R CDM methodologies they subscribe to.

Both projects are situated in the Kiboga and Mubende districts which fall into what is called the 'cattle corridor'; a belt of land where cattle grazing has traditionally been a primary economic activity, running from south-western Uganda, across the mid-west and the Nile and up to Busoga, Tesa and Soroti in the North. In particular, the Mubende–Kiboga hills in the Kiboga district constitute an important part of the Kafu and Katonga river systems. New Forests has two plantations: the first in Mubende, in the Namwasa CFR also serves as the A/R CDM project, and the second is the Luwunga CFR in Kiboga. The Kikonda CFR leased by Global Woods is also in Kiboga, near the border of the Masindi district. Both projects' baseline scenarios rely heavily on deforestation narratives, as do the plantations companies themselves, pointing to the 'encroachment' in the reserves and clearing for agriculture as responsible for the degraded state of the reserves. There is no shortage of deforestation narratives in Uganda; where it is estimated that 6,000 hectares (ha) of forest cover is lost each year (WWF, 2013). But while encroachment and agricultural clearing are prominently blamed, the loss of forests in Uganda is mostly due to a transition of forest to bush/grassland (8.2 per cent), nearly double that of transition to cropland (4.2 per cent). In fact, most woodland loss in the cattle corridor is as result of, amongst other things, overuse for subsistence grazing and commercial fuelwood (Nakakaawa *et al.*, 2010, p.49). But instead of proposing revisions of charcoal supply chains and production regimes, or the replanting of fast-growing indigenous Musizi species in CFRs suitable for charcoal production, the dominant narrative is that exotic species should be planted by plantations companies to protect 'natural forest', and carbon payments made to incentivize forest care. In this way, Leach and Mearns highlight that 'received wisdoms' and 'powerful, widely held images of environmental change and powerful assumptions about environmental crises' obscure a plurality of other possible views and policy in Africa (Leach and Mearns, 1996, p.1).

It is the re-imagining of the CFR territories as 'degraded' that scientifically justifies the development of plantations and carbon forestry (Nel and Hill, 2013; Lyons and Westoby, 2014). The powerful narrative of encroachment vilifies local people as environmental vandals and paints the private sector as saviours. In effect, the encroachment and 'illegal activities' simultaneously establish the conditions of 'degradation' that incentivize 'more productive' use of the lands, and serve as the justification for evictions.[6] Some more context is required to understand how this works. A biodiversity study conducted in the late 1990s, which led to the Nature Conservation Master Plan, tried to address land-use zoning. The plan distinguished categories of conservation and production land, with key areas such as the Namwasa and Luwunga CFRs specified as crucial sites for biodiversity conservation[7], which should remain intact, or ideally be allowed to regenerate 'natural' forest, while others such as the Kikonda CFR were set aside for 'production' purposes (initially for timber extraction). The CFRs designated as of 'Ecological and Biodiversity Importance', however, do include zoned areas where 'productive' activities (an amorphous category) can take place, but the NFA holds that intact forests within such CFRs should not be cut down or be otherwise utilized. In practice, however, these categories do not hold, for the Namwasa and Luwunga CFRs were 'pragmatically' allocated licences for plantation afforestation purposes in their 'production zones' precisely because of encroachment and its related degradation.

Similarly the designation of Kikonda as a woodland forest with encroachment allows its conversion to a monoculture plantation, rather the clearance of resident people, as well as the remaining indigenous woodlands is legitimized, raising multiple social and environmental concerns (interview, Raymond Katebaka, African Union of Conservationists, Kampala, July 2012).

It is, therefore, in the confluence of state claims to territory (and interests in timber planting) and external finance invested through timber plantation and carbon sequestration, that landscapes are recreated for 'fortress conservation' (Brockington, 2002). The territoriality of much of the protected forest estate in Central Forest Reserves is contested and CFR boundaries are perforated and unclear, with migration and encroachment following on the heels of political instability and conflict (Nel, forthcoming). Yet, because previously migrant/displaced populations are considered as 'foreigners', and are unable to prove connection to the land for more than 12 years (which would ostensibly trigger their rights to land claims under the 1975 Land Act – a clause the National Forestry and Tree Planting Act of 2003 is at pains to undermine), this again creates a conducive environment for investment in protected areas as 'degraded land', ripe for reforestation.

The controversial evictions by New Forests of so called 'foreigners' from Namwasa and Luwunga reserves is an example, even though Oxfam in a report in 2011 established that many of the affected communities had been in the area for over 40 years (Grainger and Geary, 2011). Because of state complicity in such cases, blame can be diffused by a practice of 'finger pointing', whereby projects will blame forestry or wildlife officials for the issues they experience. However, the state should not be conceptualized as a monolithic, coherent entity with a single purpose, but as an assemblage of often competing interests and agendas (Allen and Cochrane, 2010). Yet, as the experience of other carbon forestry initiatives, such as Green Resources' Bukaleba Project (Nel and Hill, 2013), as well as the Kikonda project, shows, it is not always the case that there will be support from the state for evictions. State actors can perform more subversive roles, where forest authority staff might accept bribes from 'encroachers' to allow them to plant their crops in protected and project areas. In such cases the projects still need to work with government staff and officials and represent them as reliable partners, so corrupt practices may go unreported. I will return to these 'zones of awkward engagement' below.

Disproportionate benefits and costs

In this section I set out some of the disproportionate benefits and costs borne by communities inside or adjacent to the project areas. In the Global Woods Kikonda reforestation project the community impacts have been called into question by Eklof (2012). He reported that the project cannot prove biodiversity or community benefits, or even an adequate understanding of the community numbers adjacent to the project, even ten years after the project's inception. This is despite certification with the CCBA standard, which he characterized as weak and inconsistent,[8] displaying a strong inclination among its accredited auditors towards approval, even in cases where the information presented indicated that standards had not been met (Eklof, 2012). The company did have plans for community benefits and tree-planting activities in the 'buffer zones' (set out in the CFR management plan that would bolster the integrity of the reserve) and on registered community lands. From 2005 there was to be a collaboration with 300 community members, including individual households and institutions such as the church and schools, through a group called the Kikonda Community Forestry Association (KiCoFa).[9] However, this fell away in late 2009, due to the combined effect of a lack of registered titles – held by only 4 per cent of the community (Eklof, 2012) – and the realization that planting in buffer zones might offer usage rights to communities in the buffer zone of the reserve. The company claimed that the NFA was not willing to have the land sub-licensed[10] to farmers (Peskett et al., 2011).

Income generation and employment from the project thus remains one of the only development benefits. The local councillor for Kantanabirwa village, for instance, says that shopkeepers in the villages of the area are grateful for the project, and that some are employed by it, but is critical that it will only 'accept educated people' in the office, and predominantly employs migrant labourers under poor conditions in the field (interview, Kikonda, October 2012). Migrant contract workers from various parts of the country, including Arua, Mbale and Mityana, highlighted that the contracts are temporary, on tender to middlemen who employ the migrants for 200,000 shillings per month (around US$78), depending on their outputs (Global Woods contract workers, interview, Kikonda, October 2012). There have also allegedly been incidences of rape of local women by these workers and forest rangers (CDI, 2012).

The project has had contestations with communities in its lifespan, beginning with the historical displacement of encroachers and a strict 'militaristic' management approach, whereby, since 2000, up to 100 guards policed the reserve and discouraged cattle grazing (John Mary Kisembo, Community Officer, Global Woods, interview, Kikonda, October 2012). There were further issues concerning the restriction of agricultural and grazing access amidst so-called contradictions in management, whereby at times, grazing was allowed for a fee, whilst at other times fines of 600,000 to 1 million shillings (up to US$400) were levied (CDI, 2012). Shortages of firewood, herb supply, competition for employment and erosion and loss of soil fertility on steep slopes (Peskett et al., 2011) have also been detailed, while cattle keepers lost access to the 'valley dams' that were specially constructed for cattle keepers in conjunction with Irish Aid in the reserve in 1992. After 2009, an internal company review indicated the need to change strategy, and the company has built two dams outside the reserve for cattle access. It has also worked towards better communication and a more conciliatory approach to communities in which cattle keepers are empowered to play a role – essentially to inform communities (and encroachers) of the company's short- and long-term planting plans.

The two New Forests company plantations claim sustainable community benefits from the company's plantation activities in the reserves, and from the carbon forestry project (New Forests Company, 2014). The company describes itself as a 'sustainable and socially responsible' company contributing jobs, revenue, carbon finance and the timber products the country needs as it develops, which they argue would otherwise be logged from natural forests (New Forests Company, 2011).[11] Nevertheless, allegations of 'land grabbing' and enforced, uncompensated evictions have been lodged against New Forests in a report published by Oxfam (Grainger and Geary, 2011), in conjunction with its local partner the Uganda Land Alliance (ULA). They alleged that the company enforced the eviction, executed by the government, of 22,500 people (though the number is more likely around 15,000) from these areas to make way for the forestry plantations in Kiboga and Mubende, the latter doubling as a carbon forestry project under the A/R CDM methodology (Grainger and Geary, 2011). In contrast, villager testimonies suggest that no consultations or resettlement plans were undertaken prior to eviction, and that people were violently forced from their land. Oxfam's call for transparency and fairness was certainly justified, as Human Rights Network's (HURINET) Peter Magelah reflected on his observation of the evictions: 'there were so many army and police, and the biggest problem was not involving the communities, they just gave notice then beat and chased them. They nearly arrested some people from HURINET, and some New Forest officers were very, very arrogant' (interview, Kampala, August 2012). Furthermore, this encroachment was not a one-off, isolated incident, but ongoing, and persistent. According to an NFA official the evictions between 2009 and 2011 at Luwunga, for instance, followed evictions in 1988 and 1996 (interview, September 2012).

As already highlighted, the NFA dispute Oxfam's claims, 'strongly denying' that they had any involvement in any evictions or violence, and portrayed most of the evictees as foreign, Rwandan encroachers (New Forests Company, 2011). Evictees, by contrast, state that they did have lawful entitlement to the land, some of them had lived there for more than 40 years. Others claimed to be Second World War veterans and their descendants, who claim they were allocated the land in recognition of service (REDD-Monitor, 2011). These details serve to highlight that the territoriality of these CFRs is disputed and contested, as indeed is the case with many in the country (Nel, forthcoming).

Exclusions and resistances in 'zones of awkward engagement'

Both projects' boundaries are more permeable 'on the ground' than they appear on paper, and both operate in a contested forestry context. Only one has supported evictions, while the other has come to deal with encroachment in more circuitous ways. So why have these projects had such different experiences? The answer lies, as Biddulph (2011) puts it, 'behind the facade', through the ways in which actors deal with local realities and dynamics, adopting more flexible approaches to their engagement and relations with project residents ('encroachers') and adjacent communities, than ostensibly allowed within national forestry policy and practice.

Reflecting first on the Global Woods project, the Community Officer for the local subsidiary states that they are attempting to 'learn between the lines of the law and context of the investment' (interview, Kikonda, October 2012). In essence, they are trying to work around the reality of the encroachment and crop planting that is occurring on the Kikonda CFR, in a way that does not prejudice the company's planting schedule. As he states, 'the aim is not to encourage encroachment activities, but the minimum expectation is not to hamper the expansion programme'. Officials at Kikonda Central Forest Reserve express a similar view of the management of the forest resource, which is more inclusive of community interests, and in which the line between the legal and illegal has become opaque. They allow communities to plant and harvest temporarily in the areas yet to be cleared for plantations forestry, a practice called '*taungya*', which used to be a formalized activity, but was phased out at the time of the forestry sector reforms.

The issue with these sorts of arrangements is that they constitute a tacit blurring of the legal and illegal, and further opportunities for patronage as a form of informal governance, and so further 'encroachment', which continues to be criminalized in law. The forest-adjacent communities and encroachers allege that they were sold the unplanted land in the CFR by previous Global Woods employees, and by the NFA's forest supervisor. Global Woods staff acknowledge the latter, referring to the supervisor (and former Global Woods employee) as 'the number one crook'. He and his deputy have allegedly been renting unplanted areas of the CFR to cattle keepers and accepting payments to plant land. This has caused further social conflict between cattle keepers and local agriculturalists planting in the reserves.[12] However, the company has not made a significant protest about the issue, perhaps as it aids their planting and it would not be in the company's interests to do so. A speculative assumption might be that the movement of ostensibly illegal activities further up the reserve both relieves the tensions between the company and the community that are created through the expansion of the plantation (indeed parts of the plantation have been subject to arson), and makes the planting easier, as the indigenous woodlands are being cleared.

In the New Forests project there was a very different form of engagement. Beyond preferential access, including cheap leases and tax exemptions under a government mandate for investment promotion, the NFC had political support[13] for the removal of encroachers and the enforcement of the agreement with investors, which stipulated unpopulated land in the reserve. In contrast to a number of the other carbon forestry projects in the country, including the Green Resources Bukaleba and FACE Mount Elgon reforestation (Nel and Hill, 2013), where evictions have not been allowed, connections to the president (who, it is rumoured in forestry circles, has a cousin linked to the company) were cemented after NFC and Uganda Tree Growers Association (UTGA) representatives met with the Presidential Round Table on Investment in 2009. Through the facilitation of local forestry officers, the police and the army, the company evicted encroachers from the reserves.

   To be fair to the New Forests Company, it may also have tried to be more accommodating to the encroachers who were evicted. As one expatriate working in forestry in Uganda put it, 'it's nothing to do with removing encroachers. They have wanted to help and give assistance but they were told no this is a government responsibility please wait' (Bill Farmer, Uganda Carbon Bureau, interview, June 2012). The Government rebuttal of the Oxfam report (Hon. Maria Mutagamba, Minister for Water and Environment, 2011) also mentions that it had to forbid the company from offering compensation of its own. However, despite appeals that the company was not directly involved in the evictions themselves, an NFA official in an area that includes the New Forests plantations, conceded the 'facilitation' – a word often used by local government functionaries to encapsulate the monetary or physical support they receive from donors or NGOs to accomplish their mandate – of the evictions, at least in part, by the company, while divulging the limitations of the local department:

> We can't go there every day, like in the whole of this sector there are four districts, and one pick-up [truck], you see that? But we told them [the New Forests Company] was what was needed there, drew a budget, and they facilitated us […] Because if we had to wait for the money of government (laughs) it was very difficult. Like now they [the NFA] give you a budget for one year and maybe Kikonda is not on your budget, and you are not facilitated to do any eviction there.
>
> *Interview, NFA Official, October 2012.*

The perspective of many actors 'on the ground' seems to be for 'pragmatic accommodation', rather than blanket evictions and their attendant violence and local political pressures. This would keep communities and local leaders 'on side' and, in turn, might also avoid communities 'cutting the forest at night' in retaliation. As another NFA official put it:

> The problem is you evict people and the land is not used. Like the local district politician says you are chasing people and where should they go? And yet some parts of Luwunga are not planted and it's not a forest – then how do you explain that it's a forest reserve? They say why not put people there while there are no trees.
>
> *Interview, NFA Official, July 2012.*

One alternative suggested, similar to the informal Kikonda arrangement, was that the company could provide a map and strict timetables of where they planned to plant so that communities could be told to leave in advance and, although not explicitly stated, plant elsewhere for the short-term (according to three month crop rotations) in the degraded reserve. Yet, while the NFC was active in exploiting opportunities provided in the 'zone of awkward engagement' between itself and the state, it was less successful in similar engagements with communities. These were limited to a failed attempt in 2008 to set aside two square miles for encroachers in the reserve so they could clear them off other areas. It appears that this initiative was impractical given the lack of coherent community organization, their dispersal, the lack of benefit it would give them and the lack of government direction on such initiatives.

There was significant blowback against the NFC after the evictions, showing the sorts of 'frictions' (Tsing, 2005) that can impede project operations. The types of actors who are vocal in opposition to carbon forestry – using framings of green grabbing and human rights violations – are much less restrained than the 'friendly critics' of prior forestry intervention like Collaborative Forestry Management (Li, 2007). Through online platforms, such as REDD-Monitor.org, critics are readily able to disseminate their opposition, and this vast platform greatly increases their capacity to discredit projects on various grounds, even if the projects are well intentioned. As one project actor put it, 'with carbon credits you could be the archangel Gabriel and still be accused of all sorts of things' (Bill Farmer, interview, June 2012). While, for a time, the uproar undermined shareholder confidence in the company, the contestation was initially contained through the government's backing of the NFC, and coercion in the threat to evict Oxfam from the country and to remove Uganda Land Alliance's license. Eventually a settlement was brokered by the Ombudsman of the IFC of the World Bank (a part financier of the project) in which the communities from Namwasa were to receive some compensation. The settlement, however, was limited to unspecified 'community development projects' and the residents have not been allowed to return to their land, while a second settlement for the Luwunga CFR is still pending. Oxfam did not challenge the government's actions or manage to uncover political connections to the project, carbon forestry (Oxfam America, 2008).

Conclusion

To conclude, I will reflect on these cases in relation to the three scenarios outlined in Chapter 1.

At first glance, the evidence presented certainly appears to support the 'green grabbing' scenario. What is evident from the two case studies discussed, and the social context of forestry in Uganda more broadly, is that even if projects are not directly violent in facilitating evictions, there are forms of structural violence (or what Nixon (2011) would term 'slow violence') against rural residents; who are labelled as encroachers and experience continued, long-term marginalizations that are normalized and made acceptable as the territoriality of the forest estate is reproduced across time and space. It is therefore necessary to revisit the territoriality of the forest estate more broadly, and perhaps redevelop more inclusive Community Forestry Management regimes before carbon forestry projects are embarked upon, rooted in understandings of forest resources as local commons.

In addressing the second scenario, there is less support in the case studies presented that finer attention to participation would necessarily enhance benefits for rural livelihoods. This is due to the fact that both projects have *already* attained certifications with schemes ostensibly meant to secure carbon 'co-benefits'. The New Forests plantations hold FSC accreditations, supported by international auditors, which survived the Oxfam critique, while Global Woods hold CCBA certification for the Kikonda project. Issues of violence and marginalization stemming from the portrayal of CFR residents as *de jure* 'illegal' is beyond the purview of these bodies. Insisting on more participation, inclusion or benefits without appreciating the broader social context of forestry in Uganda is not a feasible pathway to ensuring sustainable development.

By contrast, the third stance does offer more in the way of conflict mediation by embracing, as a starting point, the agonistic politics of inclusion and exclusion, and of dispute and deliberation around resource use and rural livelihoods. Market solutions should not be naïvely accepted, and neither should state-centric, managerialist visions of landscapes as mere forest reserves. In practice, compromises and negotiations must occur in 'zones of awkward engagement' such as those outlined in this chapter. In such zones, local residents, including those labelled as 'encroachers', must be seen as active participants. Recognizing realities on the ground weakens the hard-edged boundaries of the protected forest estate, and opens up opportunities for joint use at landscape scale that recognizes different opportunities, and diverse actors. Such a pragmatic approach to carbon forestry, centred on deliberation and negotiation, could afford an opportunity for addressing the beleaguered state of forestry governance in Uganda. This, however, means acknowledging the diverse claims and real politics of land use, and not expect simplistic solutions and managerial plans to work.

Notes

1   He poses two questions: do policy façades constrain agency by forcing actors constantly to reinterpret events to maintain the credibility of the policy narrative? Or do they enable practitioners to work intelligently behind the façade to produce valuable outcomes, but in ways that are more strategic and complex than allowed within the policy narrative?
2   Which were identified by USAID's Strategic Criteria for Rural Investments in Productivity (SCRIP) programme, in conjunction with the Plan for the Modernisation of Agriculture (PMA) and the NEAP, which have all established focus areas for investment in 'natural capital'.
3   Such as the Human Rights Network of Uganda (HURINET-U), the Advocates Coalition for Development and Environment (ACODE) and the National Association of Professional Environmentalists (NAPE).
4   International buyers have included conferences, electricity boards, festivals, petrol stations and transport companies.
5   Purportedly, when Ugandan Officials tried to re-negotiate over the rent for the 12,000 hectares lease of the Kikonda CFR, for instance, the Institut für Entwicklung und Umwelt officials who brokered the deal refused, intimating that the deal would be off if it was not signed by the time of their departure that night (Lohmann, 2006).
6   Anna Tsing (2005) observed this phenomenon in Indonesia.

7 The New Forests-leased Namwasa and Luwunga CFRs are categorized amongst CFRs of 'Ecological and Biodiversity Importance' which include 'CFRs whose main functions are to protect biodiversity, water catchments, riverbanks, lake shores and stabilisation of steep slopes' vital to the livelihoods of the 1.3 million people in Kiboga, Mubende and Kibaale (National Forestry Authority, 2005).

8 Its CCBA validation report identifies 'major' knowledge gaps, for example an assumption of 20 communities as opposed to the actual 44 adjacent to the project, and requests for impact assessments which were incomplete three years later (2012).

9 The group was founded and continues to be part-managed by a company employee, such that it has been criticized as 'more of a liaison group for the company than a group representing the interests of a clearly defined community' (Peskett *et al.*, 2011, p.223).

10 Kenneth Kakuru, an environmental advocate, describes that there is no law of 'traverse' which allows you to 'alienate' forestry resources from the state through registered activities (anything beyond a period of five years) on the lease (Kenneth Kakuru, interview, Kampala, October 2012).

11 I was unable to substantiate these claims at the time because of a barring of access to the reserve pending an ongoing court case between New Forests and Oxfam, and repeated stalled attempts to contact the New Forests management.

12 One encroacher, who had been in the area 13 years, perceived that the official was employed by Global Woods, alleging he had been charging 50,000 shillings (US$20) per acre for one season. He was never receipted and does not know where the money goes, and has been arrested three times over the last five years for charges of harassing cattle (interview, Kikonda, October 2012).

13 As the ULA points out, land deals are not transparent and connections to political patronage can have 'very direct benefits' (Vincent Kisembo, interview, Uganda Land Alliance, Kampala, July 2012).

6

IMPLEMENTING REDD+

Evidence from Kenya

Joanes Atela

Introduction

This chapter explores how a globally designed REDD+ project, the Kasigau project in Kenya, interplays with the local settings, perceptions, resource histories and past experiences with related projects as well as state institutions, to deliver global climate mitigation goals and local benefits. The project is the first in Africa to have sold credits under internationally accepted standards and has unfolded in a semi-arid ecosystem for almost a decade.

In terms of mitigation, the project generates credits from a protected 500,000 acre dryland forest. This translates into a validated annual emission abatement of 0.8 tCO_2e to 1.3 tCO_2e over a 20–30 year project period (Wildlife Works, 2008; Korchinsky *et al.*, 2008). These credits result from above-ground vegetation while possible below-ground credits accounts for methodological uncertainties. The project contributes to Kenya's target of avoiding emissions of 1.6 tCO_2e per year by 2030 (Government of Kenya, 2012). Kenya prioritises forest mitigation in the climate action plan because the forest sector is one of the main greenhouse gas emitters, contributing 32 per cent (19.6 tCO_2e) of total national emissions (Government of Kenya, 2012). Thus, the Kasigau project is expected to abate about 5 per cent of annual emissions from deforestation in Kenya.

The Kasigau REDD+ project is therefore a useful case study because of its international validation, its contribution to the government's climate change agenda, its engagement with local communities over a relatively long period, and its location in a vulnerable, dryland setting. The project is Africa's first REDD+ project to issue carbon credits under internationally accepted Verified Carbon Standards (VCS) and the Climate Community and Biodiversity Standard (CCBS).

Based on a policy process analysis, participatory fieldwork and document review, this chapter reveals how globally standardized REDD+ solutions become challenged by field contexts during implementation. When faced with socially heterogeneous communities, a mix of resource tenure and value systems and contradictory state policies, standardized plans have to be changed. Through adapting to local circumstances, the project has gained acceptance among the Kasigau people, reversing to some extent the long histories of exclusion from their resources by centralized state-based resource management regimes.

The chapter aims to unveil the project's interplay with the local settings, including resource histories and livelihoods, and to examine the role of project actors and the narratives that have been deployed; assess community and state engagement; and explore diverse impacts on land and livelihoods.[1]

The case study context

The project is being implemented in Taita-Taveta County of the Kenyan Costal region, 3°S of the equator and about 150 km northwest of Mombasa City. The project works in six villages with a population of 60,290 (Table 6.1). The area is classified as semi-arid with an average annual rainfall of 300–450mm per annum (Kenya Meteorological Department, 2012). The area is endowed with wildlife resources and is overlapped by the Tsavo East and West National Parks – the largest wildlife conservation area in Kenya. Yet a rapid rural appraisal and household interviews revealed that the area experiences major vulnerabilities, including water scarcity and poor land productivity (Atela *et al.*, 2014b).

A mix of ethnic groups, including Taitas, Durubas, Kambas and Somalis, live in the villages and own a diversity of livelihood assets which they use to pursue livelihoods such as small scale agriculture, livestock grazing and charcoal burning (Table 6.1). Most poor households lay claim to communal hills as their own forest resource while most relatively rich households have purchased shares in the ranches. Many poorer households practise rainfed agriculture involving the cultivation of maize and sometimes green grams. Richer households are mainly business people. Farming is practised across both dry and wet seasons but the wet season is more crucial for the productivity of crops and animal feed. Poor households who practise rainfed agriculture as their main livelihood may also pursue casual labour, food for work or charcoal burning in the nearby hills as seasonal coping strategies.

The mix of land tenure systems in the Kasigau area is crucial for the area's livelihood strategies. The land is dominated by group ranches. Group ranches are relatively expansive grasslands with both shrubs and trees. Group ranches began in the post-colonial period of the 1970s when human population, especially in arid and semi-arid lands (ASALs) rose. The increased demand for land in these ASALs reduced the capacity of the land to support large human and livestock populations. There was not enough land to be shared on an individual basis. The group ranch concept was implemented in various districts in Kenya in the mid-1960s and early 1970s and aimed at overcoming some of the problems related to sharing land resources. In the Kasigau area, the government ranches were reclassified and the land allocated to community group ranches formed as private companies such as

TABLE 6.1 Household livelihood characteristics

Asset	Attribute	Description
Natural	Geographical location	3° 33' S / 38° 45' E
	Settlement distance from the project	0.5km
	Agro ecological condition	Semi-arid
	Existing forest resources	Ranches, communal forest
	Forest management	Private, communal and trust
	Land size (acres)	1–2, 1–4, 4–10
	Land acquisition (majority)	Inheritance
	Proof of land ownership (majority)	Elders' consent and Allot. Letter
	Land ownership	Private and communal
Social	Crop yields (mean bags/acre)	2.13
	Ethnic composition	Taitas, Dorubas, Kambas, Somalis
	Main shocks (majority)	Drought
	Main coping strategy (majority)	Food for work
	Main livelihood activities	Farming and small scale business/casual labour
Human	Education level (majority)	Primary
	Household size (mean)	6
Financial	Main income source	Farming and casual labour
	Main expenditure (majority)	Food
Physical	Water access distance (km)	2–5
	Distance to the nearby road (km)	1–2

Rukinga, Taita Ranching Co. Ltd, among others. The groups were issued with ranch titles as provided for under the Land (Group Representation) Act of 1968 (Government of Kenya, 1968). The sharing was based on a defined livestock quota system which was not implemented. Individual members' benefits depended on herd size, especially the size of the breeding herd, which determined herd growth.

Each group ranch comprises between 50 and 2,500 individuals both from within and outside the local community, and these individuals hold transferrable shares to the ranches. Alongside the private group ranches exists communal land, mainly hills, that are managed by the community on behalf of the government. Individual lands, where households have settled, also exist in patches between the communal hills and the group ranches. The individual lands were initially part of the communal land that the local authority through the Chief had the powers to allocate to the landless locals and immigrants.[2] However, the process was reportedly characterized by corruption, with some politically connected people allocated very large parcels to which they acquired title deeds. For most poor immigrants and locals, however, the village elders or the Chief remain the main authority justifying their land ownership.[3] Of all these lands, dryland ranches constitute about 75 per cent of the total land area under the project, while the communal hills, trust lands and individual lands take up the rest. These dryland ranches were initially of little value to the Taitas who practised agriculture, and kept relatively small herds and flocks. Consequently, most Taitas initially informally leased their land to the pastoral Somalis. Many sold off their ranch shares to immigrants and moved to the hills for farming.

Later, other immigrants, including the Kambas and Durubas, came in to settle as squatters, mainly subsisting off burning charcoal from the dryland forest. With the charcoal business booming in response to the expanding tourist city of Mombasa, some Taitas returned to the area to engage in charcoaling for income. Today, sparse settlements of Taita, Duruba and Kamba communities live around the project boundary in communal lands, and most have no shares in the ranches.

The Kasigau project

It is in this context, with a complex history and a variegated tenure system, that the Kasigau project aims to avoid emissions by conserving a dryland forest constituting private ranches (50–2,500 members per ranch) and community land that spans 500,000 acres and is part of a corridor linking Tsavo East and Tsavo West National Parks, the two largest wildlife protection areas in Kenya. The project has engaged with communities in Taita-Taveta County since 2006, particularly supporting the community in activities aimed at reducing emissions on the protected dryland forests.

The project developer is a US-based private company, Wildlife Works. Wildlife Works has operated in the area since 1998, with specific interests in wildlife conservancies and ecotourism. Ranch shareholders signed conservation easements, committing them to transfer carbon rights to the project proponents. Community members, through local Community Based Organisations (CBOs), are engaged in activities meant to reduce pressure on the protected forests. A number of community projects are additionally funded through a trust fund, the Wildlife Works REDD+ Project Trust Fund (WWRPTF), which constitutes a third of all carbon revenue and is part of the agreed benefit sharing procedure. Protection of the forests for carbon and community engagement in the conservation activities and benefit sharing are the key project activities.

The project shares the carbon revenues generated from selling credits with international private companies such as Puma (EU and USA), Alliance Panapa Bank and Barclays Bank among others.[4] Until 2011, the project sold about 1.6 million tons of CO_2 at an average price of US$6. Out of this, about 1,373 tons were from communal hills, while the rest were from group ranches. The revenue from communal hills was invested directly into communal projects, while the remaining revenue was divided equally among ranch shareholders, communal project and project administration needs. The project is yet to sell credits from 2012–2013 and is still negotiating with additional buyers.

The share for community projects is channelled through the WWRPTF. The WWRPTF, the project staff argues, transparently institutionalizes fund disbursement to the community and eliminates tax charges on these community funds. The WWRPTF fund has channelled revenues to various community projects and in line with project proposals made by various groups in all six administrative villages surrounding the project area. However, carbon revenue generated from the communal Maungu Hill area is not part of this agreement, but is channelled directly to the community through the Maungu Hills Conservancy Association.

Intervention histories

The area has also been a subject of a long line of governmental and non-governmental interventions seeking to conserve the environment and alleviate poverty. These have had implications for the operations of the REDD+ project. As has already been highlighted, the REDD+ project area is adjacent to national parks, covering about 24,000 sq km. The parks' relationship with the local people has been subject to changing rules driven by state intervention. In the late 1970s, livelihood activities, including game hunting and wild product gathering, were banned. A village elder argued that, by contrast, the colonial sport hunting meant local people could obtain game meat and some game wardens were locals. However, in more recent times, there are no clear provisions for local benefits from parks and reserves. The Parks are currently managed by the Kenyan government through the Kenya Wildlife Service (KWS), and are a major source of national revenue derived from tourism.[5]

Other development actors include World Vision, who have worked in the area since 1999. World Vision is involved in conservation and livelihood activities, including the restoring of degraded areas, health, food for work, and local capacity building, among others. While the design of the REDD+ project differs, the experience of other NGO projects, such as World Vision's, shape perceptions in local communities about activities and benefits (Atela et al., 2014b).

Project narratives: Justifications and perceptions

The project narrative emphasizes the perceived low value of land in an area to justify improving the land value through a carbon project, and so improve incomes and reduce poverty. This element of the project narrative is the one relayed locally. Externally, the project emphasizes climate mitigation, and draws on global arguments. The project design argues that conserving the 500,000 acres of dryland forest for carbon increases the land's value beyond the limited benefits currently gained through grazing cattle and ecotourism (Wildlife Works, 2011):

> Afforestation of plantation species and agricultural activities cannot profitably be carried out in this sort of area due to a lack of water and a fragile ecosystem. Therefore we believe that we have demonstrated through our activities to attempt many different economic activities and the activities that preceded us

that there are no credible alternative economic uses for this land that could compete with the Project financially, or provide financial sustainability that would protect it from slash and burn use by the community.

Project Design Document for Phase 2 of Kasigau REDD+ project,
Korchinsky et al., 2008, p.26.

Livelihood vulnerabilities, due to poor crop yields and lack of economic activities, dominate community views during meetings aimed at disseminating the project's intentions, according to the area chief. According to project staff, managing such expectations of dramatic livelihood improvement was the greatest challenge in the early period of introducing the project to the local community. Project staff argue that the project presents a better economic opportunity for local people.

During interviews, local people lay more emphasis on how the project helps them address livelihood needs. They would always refer to the funds supporting community projects as 'carbon money', although the concept behind carbon credits and offsets remains opaque to them. At gatherings meant to inform the community of the project's aims, project staff explain that:

> We tell the community that we are trying to reduce carbon emissions by planting more trees. We tell them that charcoal burning is harmful to the environment and there is need to avoid deforestation. When trees are cut soil erosion is eminent and gullies are formed. However people mainly ask about what they will get from the project and even start requesting the project not to work with some people they think are corrupt.
>
> *Project staff, September, 2011.*

The dual narratives of improving land value and generating livelihood benefits are reflected within the state policies, including in Kenya's REDD+ Readiness Plan (Government of Kenya, 2010a) and the Climate Change Action Plan (Government of Kenya, 2012). REDD+ is seen as a route to addressing vulnerabilities and enhancing development:

> The REDD+ multi-sectoral program will look into the drivers and underlying causes of deforestation and degradation, as well as promote sustainable forest management for improved livelihoods [...]. All activities will be designed with a focus on benefits such as improving biodiversity and livelihoods of forest dependent peoples.
>
> *Government of Kenya, 2010a, p.3.*

Alongside livelihood improvements, the plans envisage some economic development from REDD+ investments. The plans specifically prioritize REDD+ as a low-carbon development pathway and an opportunity to attract international funds for climate mitigation, development, green jobs and the achievement of an industrialized nation by 2030:

Attracting international climate finance, technology and capacity building – the evidence base provided through this Action Plan can help development partners ensure that their investments align with Government of Kenya climate change priorities and that these investments are nested within Vision 2030 and Kenya's forest national planning process.

Government of Kenya, 2012, p.15.

Inevitably, the carbon project REDD+ is compared with other activities, including those of World Vision which focus on relief and handouts:

With World Vision, we have terraces on the land and some income at the end of every month. The project is very helpful in needy times especially during drought […] Yes the projects are different because the carbon project does not consider helping people during hard times like World Vision. The carbon project is good but should consider helping people in times of need.

Low-wealth female respondent, Kasigau, September, 2013.

Similarly, the experiences with governmental interventions around national parks play into perceptions of the REDD+ project. Locals see the REDD+ project as a more people-centred initiative compared to the National Parks, which have offered them no benefits.[6] This makes the project more appealing. As one elder commented: 'The carbon project is for the people and with the people while the National Parks are for the government' (Male village elder, Kasigau, September, 2013).

Project implementation processes

The implementation of the project relates to three main processes: carbon generation, community participation and benefit sharing. The project has two phases; the first is targeted to generate about 300,000 tonnes of CO_2, while about 49,000,000 tonnes are expected from the second phase (Wildlife Works, 2011). These expected emission reductions are verified through the VCS[7] and CCB standards that ensure that the project outcomes correspond with the global expectation of emissions reductions and local expectation of community participation and benefits respectively.

Community participation is required as part of the safeguard requirements under appendix 1/CP16 of the UNFCCC, and is also part of the CCB standard to which the project was validated. Under these safeguards, projects are expected to consult and involve communities in their activities, and above all spur equitable and sustainable community benefits (CCBA, 2010). Projects are also required to respect rights of representation of the local people in line with the UN Declaration on the Rights of Indigenous Peoples (UNFCCC, 2009).

The coming of the REDD+ initiative in 2006 was made known to the community through local contacts, including the area chief and leaders of various CBOs. Introducing new projects through contact persons as conduits into the community is a common approach used by projects in many parts of Kenya, and these approaches are embedded and legitimized as the accepted means of reaching many community members and gaining their attention. Specifically, public gatherings such as chiefs' *barazas* (meetings) were widely used to inform the community about the project.

Barazas are historical conduits for flows of resources and new development ideas from central government systems, and thus would always attract interest of many in the community who are expecting to benefit from any new initiative. However, *barazas* are also fora for local elites to capture new interventions. Concerns may therefore be raised on the extent to which *barazas* conform to the expected community participation in REDD+. Nonetheless, project staff argue that the *barazas* were just part of an ongoing process, because the project developer had already initiated conservation activities and other microenterprises many years prior to the REDD+ project and that, in any case, the community members are not the main carbon producers. Despite the fact that the project mainly targeted private ranches as the main source of carbon credits, fears that the project could take over communal and even individual land, execute evictions and restrict charcoal burning were common among most community members:

> At first they thought *Wazungu wamekuja kununua shamba letu na sasa tutafukuzua* (the white man has come to buy our land and now we will be chased away).
>
> *Interview with male rancher, Kasigau, September, 2011.*

These eviction fears were mostly expressed by those in the communal lands. These include some who are 'squatters' with insecure tenure. Many are resident in the nearby Maungu Hills. This area comes under the Maungu Hills Conservancy Association (MHCA), a local CBO bringing together people for joint management of the area.[8] Over time, the REDD+ project convinced the conservancy association to become part of the project. Not everyone has been happy about this decision, and fears of dispossession due to the project activities have risen. However, the project has emphasized the livelihood benefits of REDD+, and overall the fears of losing community land have been set aside, as people observe tangible benefits.

Figure 6.1 offers a schematic diagram of the main actors and activities in the project. It shows above all the complexity of the design, the multiple actors with different forms of authority and the way that carbon credits are defined. It is no surprise that local people find the project difficult to understand and misperceptions can arise, as it is organized very differently to others that they have experienced. It is neither a state project (like National Parks), although similar people are involved; it is not a NGO (like World Vision) and it is not just a private business (like the large ranchers, although whites are clearly heavily involved (see quote above)). Across this complex organization there are some important gatekeepers and brokers. The role of the chief, as the main entry point and host of the 'consultations' has already been mentioned. But there are also other committees, each of which

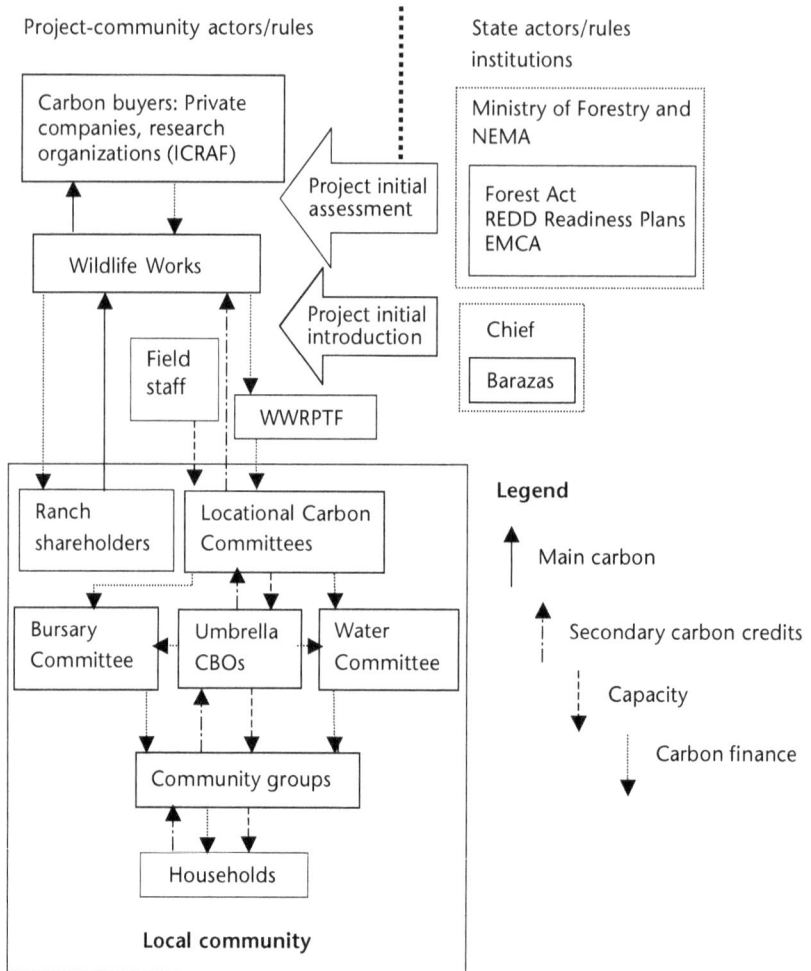

FIGURE 6.1 The organization of the Kasigau REDD+ project

have different forms of representation and accountability. And on a day-to-day basis 'the project' is essentially the Wildlife Works field staff, with whom compromises and negotiations around what the project means for whom are played out.

Effective community representation in project decisions is a key requirement in VCS and CCB standards. The new community engagement standards for REDD+ facilitated the Kasigau project to initiate new democratically elected committees to represent the community in key implementation and benefit sharing mechanisms. This has generated a general perception that the project is more consultative at implementation than other previous interventions. In practice, the project engages community members through the MHCA and other established umbrella CBOs, such as the Kasigau Conservation Trust. A number of committees, including the carbon, water and bursary committees, have also been elected by community members to represent community interests in the project's implementation decisions and activities. The committees, in liaison with the CBOs, specifically coordinate community work in sub-projects such as the eco-charcoal factory, involving the identification of charcoal-burning households and their subsequent recruitment into the factory as wage workers. They also recruit and receive funds for community forest wardens aimed to address the opportunity costs incurred by households who formerly depended on the now protected forest area. The CBOs and committees also inform and guide community members when writing small grant proposals to be funded under the WWRPTF. The project logistically, technically and administratively supports the CBOs and each of the REDD+ committees. The committees are elected by community members and often verified for gender and political representation.[9] However, while committees and CBOs form key links between the community and the project, and have been used differently by prior projects, some community members still remain outside these groups even though they benefit from the communal projects.

While creating new engagement fora is useful for the new procedures REDD+ brings, building on existing institutions and networks is advisable (Brandon and Wells, 2009). However, while these may reduce transactions costs and promote greater effectiveness, such existing institutions may be unaccountable and corrupt, and so, lack legitimacy. The project initially began working with Location Development Committees – a devolved arm of the state – however, four years later, the project delinked from these committees due to corruption, and chose to establish new Locational Carbon Committees. This has raised a question around the role of the state in the project. Today, state institutions largely remain as approval authorities of already formulated project plans, but are not part of activity design or implementation.

Land and authority: The state and land tenure

The UNFCCC has obliged the state to monitor, support and integrate sub-national REDD+ initiatives into national level reporting (decision 2/CP16) and additionally to support enabling conditions for such subnational projects (Peskett *et al.*, 2011; UNFCCC, 2012). As such, the state is the legitimate representative in REDD+ negotiations. The UNFCCC mandates states to frame REDD+ policies, readiness strategies and governance procedures. International REDD+ funds, with regard to Green Climate Funds, will also be channelled through state-based institutions, and these institutions are expected to safeguard the interests of local communities in forest resources (see Chapter 2).

In the Kasigau project, the state provides the necessary institutional and collaborative support, especially in terms of approval of processes subject to national legislative requirements. However, because national REDD+ institutional preparation for Kenya is still underway via the World Bank's Forest Carbon Partnership Facility (FCPF), the responsible state department of environment, wildlife and forestry is yet to put in place specific rules for approving subnational REDD+ projects like the Kasigau project. Nonetheless, the state subjected the Kasigau project plans to safeguard regulations contained in the Environmental Management and Coordination Act (EMCA) of 1999 (Government of Kenya, 1999). In assessing forestry-related projects like REDD+, the EMCA draws from existing forest regulations including the Forest Act of 2005 (Government of Kenya, 2005) and ongoing REDD+ readiness plans. The Act legalizes diverse forest management options, including leasehold (as in the case of the Kasigau project), public and commercial forest management. The Act allows for establishment of Community Forest Associations (CFA) as a devolved unit of forest management. These associations are constituted by groups of local people with clear interests and plans to manage forest in their areas. However, this Act does not include a legal basis for how new programmes such as REDD+ should engage local communities. It lays emphasis on how the local communities could manage or protect forests but not how they can benefit from, partner with or be protected from external programmes. Moreover, the Act does not elaborate how the state will logistically and technically support CFAs. Similarly, the Kenya's REDD+ readiness plan (Government of Kenya, 2010a) remains ambiguous on how community engagement procedures would be supported and enforced.

Subjecting a REDD+ project to state regulations, whether a Forest Act or REDD+ readiness plan, therefore may not offer comprehensive guidance, especially in respect of local community engagement. This concern is compounded by a lack of adequate capacity within Kenya's forestry sector. For instance, the Kasigau area at the time of this research had no registered community forest association. Local forestry officers argue that establishing such associations needs incentives and support and these are apparently not provided. The state is therefore likely to fail in meeting its obligation of safeguarding the local interests in forest resources that are subjected to commercial projects. In such circumstances, alternative authorities such as the chief and other local elites oversee 'participation' in resource management and allocations, inevitably shaping new initiatives in their favour. For instance, chiefs in the Kasigau area have often exercised powers over trust land, allocating it to particular households to the exclusion of certain sections of the community. Thus, in the absence of local state capacity, the project – and in this case a private company – take on state-like functions, providing services, infrastructure, extension and security in the area, as well as defining the local 'safeguarding' arrangements.

Chapter Five of Kenya's new constitution (Government of Kenya, 2010b) gave considerable attention to land reforms that envisages community participation in management of and benefits from the land-based natural resources (National Land Alliance, 2007). These reforms have important implications for REDD+ projects (Chhatre et al., 2012). As already mentioned, the project operates within a mix of private ranches and communal land. It generates much of its carbon from group ranches. However, some community members hold no share in these ranches. This segment of the community argued that recognizing the communal land in the project enables them to benefit more from the community projects. This improved their access compared to the initial arrangement whereby only individualized plots in communal areas were permitted by the chief to be linked to the project. Under the new constitution, the Land Commission's work at the local level is supported through Community Land Boards whose members are nominated by community members. The role of these boards is to coordinate land registration and inventories at local levels and report these to the commission. The Community Land Boards could provide local institutional support for REDD+ projects, yet on the other hand, the boards could compromise the extent to which land resources are decentralized and particular interests may dominate. Bureaucratic conflicts also exist between the Land Commission and line departments. For example, in the study area, the ministry of lands claims that the Commission is interfering with the work of issuing title deeds, a process that is essential for the allocation of land – and now carbon – rights. The Commission however claims that the ministry's rush to issue titles is politically motivated, with little recognition of the interests of the local people.[10]

Thus the political and bureaucratic alignment of different competing state institutions has a direct impact on what land tenure and administration arrangements prevail, with major implications for who gains access to land, trees and carbon, and who is excluded. This will have different impacts in the private ranches, in individualized plots in the communal lands and on common property resources. For example, in Kasigau there is already a move to subdivide the ranches into individual properties, abandoning the group ranch model. A single group ranch is made of multiple shareholders so subdivision would have a dramatic effect on ownership and access. There are some who argue that the individualized, land title model will improve the functioning of the REDD+ project, as clear property rights will have been defined and carbon credits can be allocated easily to individuals on the basis of land ownership. However, in Kasigau the local people have embraced a collective front and this has significantly simplified negotiations for the project.

A fragmentation of the ranches will mean that the REDD+ project will have to convince over 2,500 shareholders to commit their parcels of land to the project. This could be complex and costly, thereby reducing available funds for community projects. Additionally, disadvantaged groups such as immigrant locals, landless youths and women who own no shares in the ranches would lose the current benefits they draw through the community outreach aspects of the REDD+ project. Such exclusive land reforms are recipes for resistance, complexities and chaos for REDD+ projects, particularly because the commitment of wider community members is essential for such projects to thrive. The recent proposal to issue titles has generated some discontent among community members who claim the titles were only issued to certain sections of the community affiliated to the ruling political party.[11]

These experiences reveal that secure land terms and tenure in REDD+ ought to be viewed both in terms of ownership and in historical aspirations of a given community. In certain contexts, such as the Kasigau case, if communal ownership is well clarified and enshrined in law, a collective arrangement, with benefits that accrue to the whole community, may be the best option. The next section describes how the Kasigau project has operated within a communal land tenure system to meet the livelihood needs of the local people.

Livelihood impacts

Table 6.2 offers an overview of the impacts – perceived and recorded – of the Kasigau project. The Kasigau project therefore generates an array of impacts, shaped by history, institutional arrangements, patterns of social differentiation and resource politics on the ground.

TABLE 6.2 Perceived and actual impacts of the Kasigau Project on livelihoods; (-) negative impact, (+) positive impact, (0) no impact

Asset	Perceived impacts by households	Recorded impacts by project
	Expected impacts	Actual impact
Water access	(+) Expected construction of water projects	Ksh 3,331,551 (US$39,195) committed to community water projects
	(+) Protected water sources	
Land ownership	(+) Strengthens communal land ownership and benefits	Communal land recognized
	(-) Hinders sub-division of communal land to individual households	
Land productivity	(+) Expects rains to increase and increase yields	25,000 seedlings planted in farmers' fields
	(-) Increased number of elephants destroying crops	
Economic opportunities	(+) Diversified economic activities – from project staff and visitor	Business and employment opportunities increased (not quantified)
	(-) Restricted charcoaling/firewood collection for sale	Grazing in 400,000 acres ranches prohibited
	(-) Restricted grazing in the ranches	
Education	(+) Educational bursaries and school construction	Ksh 5,174,244 (US$60,873) committed to educate 271 secondary school students and 55 college and university students, and construct two schools
	(+) Educational bursaries	
	(0) No effect – it only targets poor families	
Employment	(+) Community members employed by the project	13 staff at the local CBO, 200 casual employees and 100 permanent employees within project activities
Household associations	(+) Maungu Hills Conservancy and associated groups supported with administrative and activity funds	
Forest cover	(+) 25,000 seedlings supplied to households	2,500 acres in communal hills conserved, over 400,000 acres in dryland forest

The prioritized interventions cut across a range of livelihood assets: natural, financial, human, physical and social. The largest share was allocated to water projects and this allocation is expected to have knock-on effects on multiple livelihood activities (Atela *et al.*, 2014a).

The impacts, both positive and negative, cut across various livelihood assets. Positive impacts include improved water access, strengthened community land ownership, diversified economic opportunities and educational facilities including bursaries. Alongside these positive impacts are negative perceptions that were more linked to social differences, especially household wealth status. Communal approaches to activities and benefit sharing was a key differentiating factor on how the poor and rich households perceived impacts.[12] In discussions, most poor people felt that the communal approach to sharing carbon benefits contributes to greater social inclusion, thereby minimising social inequality in benefit sharing. Poorer people further perceived that incorporating their communal land as part of the project improved their bargaining power for project benefits and also enabled them to benefit from carbon revenues that they would otherwise forgo with their smaller land sizes. In this communal approach every community member (whether working with the project or not, group members or non-group members, funded groups or non-funded groups) is able to draw benefits based on benefit sharing procedures that were arrived at after community consultations. Even though this might have left out non-group members from such decisions, most of these non-group members who pursue non-farm activities or non-charcoal related livelihoods (e.g. small businesses) still feel included because, first, their activities are not adversely affected by the project, and second, their inputs to the project are minimal yet they still gain advantage from project-related communal benefits. Members of a women's group involved in enhancing economic and nutritional status emphasized that water scarcity in the area affects crops yields, livestock health and agribusiness opportunities. This group, even though it was not funded as a water project, believes that a water project initiated by a neighbouring group would give them an opportunity to practise irrigated horticulture and poultry farming to supply Mombasa city, thereby expanding their economic assets:

> The REDD+ project has a greater impact than other projects because it serves the whole community and works in various lands.
>
> *Rich female respondent, Kasigau, August, 2013.*

Other than direct benefits, the project has initiated a major shift in resource governance, including creating new institutions, building capacity and encouraging greater community involvement in resource management. The newly established carbon committees charged with overseeing the implementation of community projects include elected women, youth representatives and even government representatives as members. The general feeling of inclusion among the local people is also affirmed whenever people reflect on their past experiences with centralized resource regimes, notably national parks. In other words, despite clear limitations and emerging conflicts, the project is regarded as an improvement on the past. These arrangements have enhanced the confidence of local people in their resources, something that many argue would not have happened otherwise:[13]

> The carbon project is for the people and with the people while the national parks are for the government.
>
> *Male village elder, Kasigau, September, 2013.*

While the committees and CBOs are often faced with local politics including claims of corruption and nepotism, the Kasigau people, through REDD+, have experienced a shift from a powerful state-led resource management regime to a private–public partnership regime. As such, community members feel that the project has restored their rights to forest resources, compared to before, when all benefits were channelled to the central government. REDD+ safeguarding procedures, such as obtaining carbon rights from ranches and communities through Free Prior Informed Consent, distinguish the project from state-led coercions in establishing the national parks.

While livelihood impacts were largely perceived positively, certain negative impacts were also highlighted. There were concerns, especially among the richer households, that the project is hindering the subdivision of communal land and ranches into individual pieces. This is mainly motivated by the fact that these richer households receive more carbon revenues from their shares in the ranches and are now motivated to acquire individual ownership of part of communal lands for additional carbon revenues. This was contrary to the perception of the low-wealth households, who felt that recognizing the communal land tenure system enables them to access benefits they would otherwise forgo with an individualized system because they have small parcels. Poorer households also lack legitimate titles, and feared that the rich could benefit more from subdividing the communal land due to their *de facto* financial powers. Such emerging patterns of social difference however may seed conflicts between groups. Already some conflicts are emerging between ranch shareholders and others because a third of the revenue is transferred to ranch owners in relation to their land shareholding, while a similar amount is distributed amongst a much larger group in the wider community. This gives rise to resentment among some, and clearly those with access to ranch shares are benefiting most, potentially creating a new carbon elite in the area.

Conclusions

This chapter shows that the practical implementation of a REDD+ project is faced with a host of contextual factors, intervention histories and livelihood expectations. No matter what the plans say, it is these that ultimately shape the project outcomes. In this case, issues of local authority, land tenure and institutions for participation were key features.

The success of the project was dependent on a flexible design that responded to these circumstances over time. The project offered clear livelihood opportunities through the sharing of benefits in a transparent manner, and so has been largely welcomed by local communities. Despite its limitations, the project was seen as better than the current situation on a number of fronts. The ranch areas were not generating significant incomes for land owners, shareholders or local communities, so the arrival of new carbon funds allowed an injection of resources into the area. In the past, centralized, state-led land governance had reduced opportunities for local people to benefit from their local resources. The REDD+ project by contrast was explicitly premised on benefit sharing. The provisions under the new Kenyan constitution, including more decentralized approaches to land issues, has created new opportunities for local land management institutions. The REDD+ project, driven by a private–community alliance, based on sharing benefits generated, has benefited from these new arrangements, developing new approaches to resource governance in the area. Community benefits are significant, amounting to one third of the total revenue, and these support livelihoods in a range of ways. This helps to guarantee that the carbon sequestration benefits are genuine and for the long-term.

There are, of course, questions raised about who benefits within communities, and between communities, ranchers and the project, and the longer term land tenure and governance issues remain, with proposed shifts in tenure likely to change land control and access to benefits. However, for the time being, the project, partly because it operates in an area where little other opportunity exists, and where past top-down interventions have failed local people, is seen positively by local people as well as the company involved.

Notes

1 More detail on the study's conceptual framework, methods and findings can be found in Atela (2013).
2 Discussion with village elders, Kasigau, October, 2013.
3 Household interviews, Kasigau, October, 2013.
4 Interview with project staff, Kasigau, October, 2013.
5 Interview with KWS staff, March, 2012.
6 Focus group discussion, Kasigau, September 2013.
7 The VCS approved methodology 'VM0009 Methodology for Avoided Mosaic Deforestation of Tropical Forests V1-0' is applied in carbon accounting. The method was developed and is maintained by the project developer Wildlife Works (2011).
8 Interview, representative, Maungu Hills Conservancy, Kasigau, September, 2014.
9 Interview with chair of a location carbon committee, Maungu, September, 2013.
10 http://k24tv.co.ke/?p=22803, accessed 28 September 2014.
11 www.standardmedia.co.ke/?articleID=2000092862&story_title=issuance-of-title-deeds-should-not-be-politicised, accessed 28 September 2014.
12 Household interviews, Kasigau, September, 2013.
13 Focus group discussion, Kasigau, October, 2013.

7

CARBON PROJECTS AND COMMUNITIES

Dynamic encounters in Zambia

Guni Mickels-Kokwe and Misael Kokwe

Introduction

Endowed with vast *miombo* woodlands and an alleged very high deforestation rate – 280,000 hectares of forest degraded per annum (ILUA, 2008) – Zambia is perceived as a serious contender on the international emissions markets. In 2010, the country was selected as a pilot for the UN Reduced Emissions from Deforestation and Degradation (UN-REDD) Programme, and, since then, half a dozen initiatives have presented themselves as the field-level forerunners to national REDD+ carbon engagement. It is anticipated that Zambia will soon be trading carbon offsets on the voluntary market, and that the validation of the draft national REDD+ strategy will be completed.

Strong community engagement is considered key to the success of the REDD+ process. Indeed, the draft Forest Policy of 2009 emphasizes the role of community involvement in sustainable forest management in Zambia (GRZ, 2009a). The goal of the Cancun Agreement on REDD+ safeguards is transparent and effective forest governance that respects the knowledge and rights of local communities, and allows for their full and effective participation (Matakala *et al.*, 2014). In this context, what are the first experiences from Zambia in the encounter between carbon projects and local communities?

Community participation in Zambia's early carbon engagement is examined through the lens of the Lower Zambezi REDD+ Project (LZRP) in Rufunsa District, Lusaka Province. A qualitative study of project dynamics was conducted over the period 2012–2014, based on document review, field visits and key informant interviews. The chapter focuses on the processes of engagement, local governance and the incentives for community participation in the LZRP. Against the backdrop of a heterogeneous landscape, a long history of unresolved local natural resource governance conflicts in rural Zambia, and a long-recurring projectized approach, the chapter reveals some questionable rationales – will chicken-rearing, for instance, really stop deforestation? And can project approaches really be sustained over the long timeframes of carbon initiatives?

REDD+ in Zambia

Since 2010, the UN-REDD Programme has provided the key pathway to drafting the national architecture for emissions management and carbon offset initiatives in Zambia, building on earlier environment, climate change and forestry policies. The current readiness stage encompasses situation analyses, capacity-building and stakeholder consultations for the development of a national strategy. The national REDD+ strategy will contain recommendations for policy reform, investment guidance, and an implementation framework, with due monitoring and safeguard systems, in line with the requirements of the United Nations Framework Convention on Climate Change (UNFCCC) (Matakala et al., 2014). The draft strategy is used as a proxy for the analysis in this chapter. Two key aspects emerge from the national consensus on the REDD+ strategy development process. First, that the focus will be on sustainable land use and management, encompassing all natural resources rather than forests alone. Second, that Zambia will amend existing laws and policies to implement the REDD+ process, rather than have a stand-alone REDD+ policy and legislative framework.

Zambia's vision for community engagement in REDD+

Zambia's national vision emphasizes the effective involvement of stakeholders and local communities in sharing the benefit from all forms of carbon engagement. Community involvement in REDD+ is multi-jurisdictional and multi-sectoral, reflecting the diverse proximate and underlying drivers of deforestation and forest degradation (Matakala et al., 2014).

Within the forestry sector, it is envisaged that communities will engage through existing forest management models, to be adapted to include a REDD+ compatible carbon trading framework, including Joint Forest Management (JFM), community forests, private forest management, and forest product certification schemes. Planned legislation and policy reform will create new rights and responsibilities as well as new rights-holders and interest groups. The accompanying shifts in power will affect not only carbon ownership and use, but more broadly, land tenure, natural resource use and management, participatory decision-making, benefit distribution, and the incentives that will substitute forest services with low carbon emissions.

The process is not easy. The national UN-REDD secretariat attests to the difficulty of delivering a coherent national strategy under the complex methodological prescriptions set by a vacillating international carbon trading architecture. For now, the system is far too complex for communities to engage without the middlemen, the carbon brokers (Mickels-Kokwe et al., forthcoming).

REDD+ instruments for community engagement in Zambia

In the emerging national REDD+ strategy, three aspects are particularly pertinent for the process of engaging communities in carbon offset initiatives: social safeguards, local governance and institutional arrangements.

As party to the UNFCCC, Zambia is obliged to take on national interpretation of the seven social and environmental safeguards adopted at Conference of Parties (COP) 16 in Cancun in 2010 (Matakala *et al.*, 2014). These safeguards aim both to prevent and mitigate harm from investment or development actions, and to operationalize the rights of communities, incentivize the protection and conservation of natural ecosystems, and enhance other social and environmental benefits (Rivera, 2014). Zambia's draft REDD+ strategy recognizes that the 2011 Environmental Management Act contains basic provisions for environmental safeguards, although often weakly enforced; hence, to enhance social and economic rights, new legislation will be needed. There is also a need to integrate mechanisms for operationalizing safeguards – such as Free Prior Informed Consent (FPIC) – into existing sector policies and legislation.

A key governance issue at local level is the construction of a REDD+ implementation structure that resists local elite capture and prevents external partners dominating the benefit sharing regime. Zambia's Decentralization Implementation Plan (GRZ, 2009b) already emphasizes community rights and participation. REDD+ needs to align with and reinforce this, towards the end goal of devolving responsibilities and rights to the level of forest-dependent communities. REDD+ in Zambia will succeed if communities receive tangible, sustained benefits for the duration of the carbon initiatives in which they are directly involved.

The proposed mechanism for community involvement in REDD+ initiatives are the Area Development Committees (ADC). These are envisaged as providing a network to link and share information between villages and their local forest management structures (e.g. JFM) through district councils to national government.[1]

At community level, three of the Cancun safeguards[2] will be particularly challenging to interpret and practice in the Zambian context. The requirement for '*Transparent and effective national forest governance structures, taking into account national legislation and sovereignty*', will be difficult in the context of the weak institutional arrangements for natural resource governance at local level. The requirement for '*The full and effective participation of relevant stakeholders, in particular indigenous peoples and local communities*' raises questions about what constitutes consent and participation where not all members of a diverse community benefit from REDD+ activities, but opportunity costs will affect everyone. And the requirement to take '*Actions to reduce displacement of emissions*' recognizes that the benefits from carbon performance-based incentives take time to accrue, but how, in the meantime, will poor communities be encouraged to implement long-term emission reduction activities?

These potential conflicts in implementation are envisaged as to be avoided and if necessary resolved through clear REDD+ principles, and capacity-building to apply them. However, in practice, whether this happens depends on the success of the national REDD+ aspirations, and the recognition of landscape and governance history in local carbon engagements.

Landscape heterogeneity and local history

Since 2010, the UN-REDD Programme in Zambia has spawned consultancies, project proposals, companies, NGOs and advisory groups, all of whom are aiming to capture the new value of carbon. In addition to forest conservation for carbon trading, the initiatives include agroforestry (e.g. Trees on Farms), market-oriented conservation agriculture (e.g. Community Markets for Conservation, COMACO) and low-cost energy-efficient cooking stoves (e.g. the Three Keys Project). The initiatives are beginning to have an impact on landscapes and livelihoods, with important consequences for land use, local governance, markets and the relationship between rural communities and the state. As, typically, carbon engagements are planned for 30 years, their impacts will last a long time – approximately the life-span of one generation. Casting the searchlight backwards in time, what other long-term processes influence the lives and livelihoods of the new 'carbon generation' in rural Zambia?

Landscape heterogeneity

African agrarian landscapes preserve a historical record of the relations between resource users and their environment. Over the last three generations, the study area has changed from a very sparsely settled savanna woodland, where traditional farming created only a minor impact, into a heterogeneous landscape, with a variety of rural economies coexisting, utilizing the land at very different levels of intensity. Visible in the landscape are the remnants of, among other things, early colonial decisions on administrative policies and infrastructure; the formation of the Soli Shamifwe Native Reserve in 1929; labour migration to the mines; the separation of traditional and state authority through the abolishment of the Native Authorities in 1965; the smallholder modernization project by the independent nationalist state; the change in land use in response to the economic collapse following the 1973 oil crisis and slump of copper prices; the Zimbabwe liberation war 1974–1980; market liberalization in 1991, and subsequent privatization policies; the devastating droughts of 1992–1993 and 1995–1996; the 2004 surge in international mineral commodity markets; the new infrastructure-based populist politics; and rapid recent economic growth and globalization (Mickels-Kokwe *et al.*, forthcoming). Carbon projects, as we shall see, have added to and interlayered with these embedded landscape histories.

The study area

The study area is located approximately 120km east of Lusaka, in the hill and escarpment zone to the southeast of Chongwe town. The area is a transition zone between the agriculturally more productive plateau and the harsh, dry lands of the Zambezi valley. Annual rainfall averages 800mm on the edge of the plateau. The soils are infertile, acid, shallow and gravelly, sustaining slow-growing *miombo* woodlands on the escarpment. The area is heavily infested by tsetse fly (*Glossina morsitans*) in the vicinity of the Lower Zambezi National Park.

The owners of the land are the Soli people. The Soli Shamifwe Royal Establishment resides at Chinyunyu on the Great East Road. Waves of migration have swept the area over the years, and the Soli now constitute less than two thirds of the population. The Chikunda and Ngoni moved to the Chongwe plateau in search of land in the 1950s and 1960s. In 1980, as the Zimbabwe liberation war ended and the freedom fighters left the Zambezi valley, people started moving into the area in great numbers. These include the Tonga, who arrived in the wake of the 1992–1993 and 1995–1996 droughts and locally constitute up to a third of the population, the Bemba and Lala, and even Kaonde and Luvale from the north. Village headmen particularly complain of the Mwachusa, known for roaming the land in search of work, asking for land to farm, but staying only a few years, burning charcoal and, when the trees are exhausted, moving on.

Land and livelihoods

Land is held under customary tenure, vested in the Chief, with a few private leasehold farms found in pockets of more fertile soils. Land is a contested matter, clad in layers of history. In 1929, the colonial alienation of fertile plateau land for settler agriculture along the line-of-rail resulted in the forced movement of people into the Shamifwe Native Reserve. This quickly rendered the traditional *chitemene* shifting cultivation system ineffective, causing land and food shortages from the mid-1930s (Wood, 1986).

Despite corrective administrative measures in 1942, and subsequent out-migration of young men to the mines, the land problem persisted. After the economic collapse in 1973, out-migration to the mines stalled and, as the rural population grew, more marginal land was brought into cultivation. A 1984 government report proposed subdivision of state land as a 'necessary measure' to accommodate the rapidly growing population (GRZ, 1984, p.10). Since then, the land issue has been discussed many times, but not addressed.

In the wake of economic restructuring in the mid-1990s, migrants from nearby urban areas further added to land pressure in the former reserves. The 1995 Land Act allowed for the conversion of customary land to leasehold tenure, further reducing the land available. Squatter encroachment testifies to the shortage of land. Approximately one fifth of the national park is encroached by cultivation and a portion of the Rufunsa conservancy was so encroached by charcoal producers that it had to be excluded from the carbon project (BCP, 2013). Traditional authorities are often unable to control squatters who originate from outside the chiefdom (Mickels-Kokwe *et al.*, forthcoming).

Livelihoods are composite, as in most rural areas in Zambia. Farming constitutes the mainstay, but in the poor soils of the hills and escarpment zone, household income is complemented by a range of other economic activities. Cattle are an important part of farming. A comparison of crops grown a generation apart, in 1982–1993 and in 2012–2013, shows the disappearance of shifting cultivation crops (millet) from the farming system and the introduction of new crops, e.g. vegetable farming ('gardening') (ARPT, 1984; Mickels-Kokwe *et al.*, forthcoming). An intense government effort supported by donor funding to agricultural extension under the 1980s 'Lima Ladder' programme, combined with heavy subsidies to all stages of the maize value chain, provided the rationale for smallholder adoption of maize as the main staple and cash crop (Mickels-Kokwe, 1998). Despite the bio-physical constraints in the area, agriculture is still considered the main driver of development.

Interestingly, the most important off-farm sources of income have remained the same over the past 30 years: beer brewing for women and charcoal production for men, though petty trading has become an important addition. The income earned off-farm is small and poverty is prevalent (BCP, 2013). Eighty per cent of households are considered poor, a condition that fellow villagers often consider self-inflicted by laziness and drinking. Hard work is the ladder out of poverty. Maize farming and cattle are considered the means to success. The ultimate indicator of wealth is the ability to diversify – into shop-keeping, milling or the transport business (Mickels-Kokwe *et al.*, forthcoming).

The city of Lusaka exerts a major influence on the study area, as a source of employment and market opportunities. Commercial charcoaling for the Lusaka market commenced around 1970 in the forests near the Great East Road (Chidumayo, 2001). The practice has spread, reaching the National Park and beyond, and has been intensified by the recent rehabilitation of the two main feeder roads (BCP, 2013). The good agricultural and livestock potential and access to markets of the nearby Chongwe plateau attracts investment in farming. At the same time, Lusaka acts as a drain on labour, provides competition and pushes up casual farm wages.

A history of weak local government and project-focused development

The search for optimal natural resource governance structures and decentralized local government remains an unresolved issue in Zambia. While the REDD+ draft strategy proposes the adoption of Area Development Committees as units of community involvement, the ADC is a relatively new institution.

At Independence in 1964, Zambia inherited a system of colonial administration, whereby local government at sub-district level worked through Native Authorities, headed by Chiefs. In 1965, the Native Authorities were abolished, regarded as remnants of colonial oppression, and were soon replaced by district councils. The new state was also very conscious of the need to create a nation transcending the divides of ethnic groups (Lungu, 1985). Chiefs remained marginalized until 2012, when debate resumed on how to effectively engage them in local governance.

Finding an administrative space between the village and the district was difficult. Attempts to introduce wards – 'an administrative unit consisting of a number of villages' (Waern, 1984, p.1) – were unsuccessful for a variety of reasons (Lungu, 1985). From 1980 to 1990, during the era of the 'One-Party-State', wards merged with the ruling United Independence Party structure. This sidelined their administrative and developmental functions (Waern, 1984, p.2) and meant that community participation became restricted to only members of the ruling party.

To compensate for the lack of non-political sub-district structures, many technical ministries and donor projects established their own parallel 'committees', such as for water and sanitation, health and agriculture. After the economic collapse in the mid-1970s, the Zambian state became increasingly dependent on external support (Wood, 1986) and the 'project approach' gradually became mainstream development policy. Past projects in the study area range from cotton schemes and tractor hire, to participatory village development, food security, outgrower schemes, income generation and conservation agriculture. Local people assess carbon initiatives through this historical frame of reference.

Decentralized local governance structures did not gain legitimacy and capacity through the 1990s in the context of multi-party democracy, structural adjustments and cuts in staff and resources. As electoral units lacking infrastructure, the wards were sidelined in the rapid expansion of development aid and its proliferation of committees, resulting in severe compartmentalization of rural affairs into separate 'silos'. In 1995 the government introduced the Provincial and District Development Coordination Committees (DDCCs), and then District Planners, in an attempt to coordinate activities.

It was from 2003, amidst a renewed attempt to resolve the governance vacuum at sub-district level through a Decentralization Policy and Implementation Plan (DIP), that the ADCs were proposed (GRZ, 2009b, p.31). ADCs are to engage in resource mobilization, revenue collection, community sensitization, prioritization of projects for inclusion into the district strategic development plans, monitoring and evaluation of implementation activities. Early experience suggests that the ADCs may be effective in coalescing community participation, coordinating the lower levels, and gaining community confidence (Mumba, 2013).

The DIP also revitalizes the roles of Traditional Authorities (Chiefs and their royal establishments), recognizing their roles in community mobilization and as resource custodians (GRZ, 2009b, p.31). With the coming of the Patriotic Front government in 2011, the role of the Chiefs is being revisited through the establishment of the Ministry of Chiefs and Traditional Affairs and proposed legislation for the integration of traditional authorities with local government.

While the new carbon engagement views community participation as essential, then, this is in the context of a history of weak local governance and very limited community involvement in local affairs – including natural resource management. The current decentralization push offers potential to enhance community empowerment and resource control, but history creates challenges for meeting the goals of effective governance, participation and incentives for long-term conservation. Moreover the current adult generation in rural Zambia, aged 20–50, were born into the local governance vacuum; they have seen no other mode of rural development than an incapacitated local government, and the proliferation of seemingly random projects.

The narratives underpinning the rhetoric of project-based development have focused on the role of individual resource users as agents of resource degradation and deforestation, while ignorance and all-encompassing poverty provide explanations for local failure to develop economically (Mickels-Kokwe *et al.*, forthcoming). Project development has fostered a pervasive belief in the necessity of externally driven modernization to lift rural areas out of their misery. Rarely are the underlying structural causes addressed, such as the persistent government failure to address politically difficult issues of land shortage and local governance. Carbon initiatives are the latest in this lineage of iconic projects, now executed in privatized, commoditized and globalized styles of intervention.

The Lower Zambezi REDD+ Project[3]

'Ba Carbon. They are the carbon people'.[4]

The Lower Zambezi REDD+ Project (LZRP) is an ambitious initiative marketed as an innovative, unique and transformative project design, 'intended to sustainably improve local livelihoods and provide viable alternatives to activities that are dependent on deforestation' (BCP, 2013, p.193). Progress is rapid. In 2013, the project proposed to 'pilot' a new methodology for community engagement (BCP, 2013, p.63). By 2014, the LZRP was already ready to 'serve as a model for the implementation of additional community-based REDD+ projects in Zambia'.[5] Yet given the historical context of slow social change in the greater Chongwe area, the substantial claims made by the LZRP project are intriguing. Will the carbon project succeed where others have failed?

The LZRP, developed and managed by BioCarbon Partners, commenced in 2009 and is 'the first REDD+ pilot project in Zambia'.[6] The project aims to reduce emissions from deforestation and degradation on 38,781 hectares of privately owned land in Rufunsa District, known as 'Rufunsa Conservancy'. The project was developed to conform to the Climate, Community and Biodiversity Standard (CCB Standard, second edition) and the Verified Carbon Standard (VCS, version 3.3). The planned emissions reductions are 9,615,646 tCO^2e, over 30 years. The project was validated in 2013.

The LZRP has two components, a conservation and a community component. The Rufunsa Conservancy is managed for forest and wildlife conservation, and the estate forest is the source of traded carbon offsets. The benefits from carbon trading accrue to the two parties to the 'Carbon Rights Covenant': the project developer and the land owner (BCP, 2013, p.129).

As the LZRP was developed to conform to CCB Standard, it has a strong community component. The project engages the surrounding communities in project activities intended to mitigate deforestation, which provide incentives to local communities to reduce harmful practices and change behaviours that contribute to forest degradation. The community component of the LZRP engagement is the focus of this chapter.

Institutions, actors and visions in project development

BioCarbon Partners (BCP) is a private company registered in Mauritius, with strong personal links with Wildlife Works, and experience in Kenya (see Atela, chapter 6, this book). BCP is described as 'an African-headquartered, focused and majority African-owned social enterprise that develops and manages long-term forest carbon projects in globally significant biodiversity landscapes in Africa'.[7] Two subsidiary structures have been formed in Zambia: BioCarbon Partners Limited provides employment for the project staff; and BioCarbon Partners Trust, a not-for-profit organization, helps leverage donor funding and assists with the design, implementation and management of community projects (BCP, 2013).

BCP implements the LZRP in close collaboration with immediate partners, namely the owners of the Rufunsa Conservancy, Sable Transport Limited; the traditional authority, Chief Unda Unda of the Soli Shamifwi Royal Establishment; and MUSIKA, a Zambian NGO providing community mobilization and engagement support. Several organizations have been engaged in specific project activities, e.g. the Conservation Farming Unit and Engineers Without Borders. Recognition by the Zambian Government has been obtained through letters of support from the Department of Forestry and District Council (BCP, 2013). The backdrop is a vast network of advisors and international NGOs, and a technical advisory board.

BCP considers LZRP 'a discrete project with high potential value in the marketplace given its anticipated climate, community and biodiversity benefits'.[8] Initial capital for the start-up phase was obtained through private investors and financial institutions. Very soon, however, BCP embarked upon leveraging funds from cooperating partners to accommodate the costs of community engagement. During the project design period, funds were obtained from organizations such as the Civil Society Environment Fund (CSEF, funded by Denmark and Finland), the Department for International Development (DFID) and the United Nations Development Programme (UNDP) (BCP, 2013).

The BCP operations in Zambia are not confined to one project area. In early 2014, BCP was selected to implement a vast 700,000 hectares USAID-funded community forestry initiative in the nearby Eastern Province, through which BCP will gain carbon emission trading rights.[9] A new investment partnership was formed with the BioCarbon Group, a global investor in land-based carbon mitigation activities with shareholders including Global Forest Partners LP, the International Finance Corporation (IFC), a member of the World Bank Group and Macquarie Bank of Australia. BCP has signed a multi-year Investment Agreement with the BioCarbon Group to support the LZRP, which BCP is implementing. Funding from this investment will be directed towards strengthening community livelihood activities, forest and biodiversity protection, carbon verification activities, credit marketing, and collaboration with Government.[10]

BCP's vision is to 'achieve transformational social development and conservation returns in Africa through REDD+' (BCP, 2013, p.59). Relating to the community component, the specific objectives are the creation of employment opportunities; the alleviation of poverty through sustainable natural resource-based and other economic activities; support to the provision of critical social services; facilitating market access and linkages; and piloting community covenants as a tool to link community activities to the mitigation of deforestation and biodiversity threats (BCP, 2013).

In the long-term, the LZRP is intended to become self-sustaining. The business model for the conservancy component is built on a combination of eco-tourism and game ranching. The deforestation mitigation activities in surrounding communities are, in the words of the Project Design Document, 'designed from the outset to become self-sustainable' (BCP, 2013, p.88). The anticipated impacts of project activities on local community well-being by 2039 are all positive, qualitative indications of improved income at household level to broader community level indicators such as 'sustainable use of forest resources' (BCP, 2013, pp.140–143).

Local project experiences

The previous sections have outlined some of the expectations and key challenges for community participation in carbon offset projects. This section analyses actual project encounters in the LZRP through the eyes of stakeholders and local communities.[11]

Local governance – institutional anchorage

BCP frequently interacts with the UN-REDD secretariat at the Forestry Depart-ment (FD) in a collegial and consultative manner. The start-up timetables of projects coincided and the sharing of lessons learnt appears to have constructively informed both processes. Beyond the REDD+ context however, the institutional anchoring of the LZRP is less apparent. What can the project teach Zambian institutions hosting future carbon offset projects?

According to the Project Design Document, BCP sought operational approval from Zambian authorities, among them the FD and the Chongwe District Council through 'letters of support' (BCP, 2013). The need for more firm collaboration quickly arose, but the process stalled as the then Director of Forestry requested greater detail of the anticipated benefit sharing mechanism. By July 2014, the district and sub-district level FD staff had still not seen any written agreement. The lack of clear rules of engagement may account for the erratic collaboration observed, in which FD staff at times appeared reluctant to commit scarce resources and their time to LZRP activities (Mickels-Kokwe et al., forthcoming).

At district level, BCP sought recognition in the Offices of the District Commissioners (DC), the political heads of local government in Chongwe and Rufunsa districts. The project developer regularly visits the DCs' offices. The DCs in turn would attend key project events, such as the project launch and consultation meetings. Beyond the DCs, however, the project was not well known in the district. The District Agriculture Coordinators (DACO) and the District Forest Officers (DFO) were either not aware of, or only held superficial understanding of the project (Mickels-Kokwe *et al.*, forthcoming). The working relationship with Zambia Wildlife Authority (ZAWA) was more intense.

Quite significantly, we found no platform for regular coordination with the district authorities. The LZRP does not attend DDCC meetings, nor are there procedures for regular consultation with any of the relevant sub-committees or departments. Indeed, the ward counsellor expressed great disappointment at the distance kept by the LZRP field teams (Mickels-Kokwe *et al.*, forthcoming).

Given the limited visibility of the LZRP, questions of transparency, accountability and collaboration arise. How does the ordinary public gain information about a project that is not known in the district, only advertised on the internet? How does the project developer ensure that their project activities support those of other actors in the district if consultations are sporadic? It appears that the LZRP strategy is based on personalized relations rather than robust anchoring in local government institutions, threatening sustainability in view of the high turnover of staff in both BCP and the Zambian government bureaucracy.

Community engagement – strategies and practices

'They hold meetings without tangible outcomes in terms of activities'.[12]

The LZRP undertakes conservation activities in the Rufunsa Conservancy, and deforestation mitigation activities in surrounding communities. The catchment area comprises communities currently involved in charcoaling or agriculture, within 30km of the Conservancy, a total of 8,300 people living in 28 villages divided into four zones. These communities are 'therefore reasonably expected to be impacted by the implementation of REDD-related activities on Rufunsa conservancy' (BCP, 2013, pp.9, 16, 80, 88). It is not clear from the documentation whether the communities were presented with the option of *not* participating in the LZRP.

BCP's approach to the process of community mobilization comes across as positive and passionate. Community engagement is described as:

> [U]niquely comprehensive in its involvement and empowerment of local communities throughout the decision-making process concerning REDD+ project activities. The strength of our model lies in the way that we treat local communities as 'allies' [...] By thoroughly consulting and involving

communities [...] we are able to benefit from local knowledge and legitimacy that could not come from less inclusive processes [...] we believe our project activities will be more robust and effective if they have been designed and supported by local communities. We are proud of the fact that we can truly describe our model as 'community based'.

BCP, 2013, p.89.

Building collaboration with more than a thousand households scattered over an area of 1200km^2 in two years is no easy undertaking. Is the community engagement as strong as this in the eyes of community members? The BCP community engagement strategy rests on several steps (BCP, 2013). First, sensitization meetings were held to inform communities about the objectives of the LZRP. These included basic information about climate change and REDD+, in accordance with the principle of obtaining FPIC. A commendable estimated 58 per cent 'sensitisation rate' was achieved (BCP, 2013, p.91). Subsequent steps, community identification of potential projects and formation of Zone Development Committees, are discussed in the sections below. A baseline survey conducted in September 2012 was the fourth step, involving interviews with 90 households. Finally, a field-based Community Engagement Team was formed to maintain ongoing dialogue with the communities, stakeholders and traditional authorities.

Undoubtedly, BCP has invested exceptionally in the process of mobilizing and engaging communities. However, best practice from rural development initiatives in Zambia clearly demonstrates that a still greater effort is needed to sustain the momentum generated in the first sensitization than is described in the BCP documentation. Past projects suggest that an average attendance at five to seven meetings per individual is needed for rural inhabitants, with limited education and almost no access to newspapers and other media, to gain a level of confidence that allows them to take decisions and adopt new practices (Mickels-Kokwe *et al.*, forthcoming). This is reflected in our interviews. Headmen, who have attended six or seven LZRP meetings, eloquently explain what carbon is and how the project operates. Other members in the communities, who have only been present at one or two meetings, are hesitant and lack confidence in their answers. They may remember what was said about climate change in the sensitization, but do not understand the project concept well. Creating awareness and providing basic information is therefore only the first step. Building effective community commitment requires repeated dialogue and deeper partnership-building, especially where complicated concepts and contractual arrangements are involved. For example, we observed that, while a few could explain concepts such as carbon as a product, no one could explain the concept of carbon *trading*, nor the relationship between BCP and Sable Transport, the owner of the Conservancy, suggesting that this aspect of BCP activities had not been discussed in community meetings.

Local governance – sub-district structures

Once community engagement started, BCP quickly discovered the institutional void characteristic of sub-district governance structures. At village level, an existing structure was found: the traditional Village Committee comprising of the headman and his advisers. Between the village and the district they encountered a governance vacuum. There was no suitable platform for community members to participate in project matters. To remedy this, BCP created a new governance structure, the Zone Development Committee (ZDC). Alignment with traditional leadership structures was perceived to provide desired local legitimacy and approval (BCP, 2013). The documentation does not reveal what other, if any, alternative institutional arrangements were considered.

BCP envisages the ZDC as a 'transparent forum to filter eligible community project opportunities that are appropriate across village boundaries and share information of village project progress' (BCP, 2013, p.32). The ZDCs are described as innovative, 'hybridized decision-making bodies that build upon local leadership structures, while introducing new elements of democratic representation and intra-zone cooperation'. Further, the ZDCs are expected to serve the function of being 'a form of ongoing participatory rural appraisal system (PRA); as a forum for local people to examine their own problems, set their own goals, and monitor their achievements' (BCP, 2013, p.96). The newly formed committees are intended to assist BCP in identifying projects, project models and 'selecting the most appropriate individuals or groups' (BCP, 2013, p.99).

In early 2013, LRZP facilitated the formation of four ZDCs. Each comprises two democratically elected representatives from each village, in addition to the representatives from the Chief's Village Committees. Each ZDC therefore has at least 20 members, a rather large number of people to mobilize for a meeting. The project document is silent on the modalities for how the ZDCs meet and whether they conduct anything other than LZRP business.

In a broader context, the solution seems a repeat of the stand-alone development committees of the 1980s, described above. The ZDCs, to the extent they are understood by the local authorities, maintain a 'project' status in the district. But is it desirable for an initiative that will last for 30 years to operate as a 'project' throughout? Without institutionalizing the carbon initiative into the sub-district administration in some form, it carries little recognition or oversight from local government. Currently, the ZDCs are not recognized in forest management or as REDD+ operational structures either. In the longer term, such stand-alone structures do not appear to be viable solutions. Moreover, in the context of multiple REDD+ interventions in the same area, should they share the same committee or start their own? Numerous questions arise on the legitimacy, mandate and authority of the committees to take decisions on anything beyond carbon project business. And building the capacity of stand-alone project committees is expensive, threatening to become a financial burden on the carbon project developer in the long run.

Community covenants

BCP's overall community-based model places partnerships with communities at the heart of forest protection. Rather than pursuing an enforcement-based model of land-management, BCP hopes to invest heavily in communities so as to meaningfully and sustainably address the drivers of deforestation. By allowing communities to feel benefits derived from cooperation towards forest protection efforts, BCP hopes to shift incentive structures away from unsustainable deforestation, towards more sustainable (and profitable) livelihoods. In so doing, BCP allows local communities to serve as critical partners or allies in the implementation of REDD+ activities (BCP, 2013). How is this partnership to be structured?

The community covenant, a signed agreement between BCP and community representatives, is the basis for the partnership. The community participants commit themselves to reducing their non-REDD compatible activities in exchange for project investment, support or employment. BCP's community covenants are designed to serve as mutually binding 'contracts' that link project activities and community interventions with deforestation mitigation and biodiversity enhancement efforts, including reducing the risk of 'leakage' (BCP, 2013, p.63).

Community covenants underpin every single community activity, explicitly documenting that BCP support is conditional upon tangible progress towards reducing deforestation. Thus, participants in the sustainable eco-charcoal community project have signed community covenants that document their agreement to stop unsustainable practices in exchange for the benefits and support. The community covenant contains a clause on FPIC, for example:

> By signing below, the participants in this project confirm that they have been fully informed about the terms of their participation in BCP's Sustainable Eco-Charcoal Project, and that they have freely provided their consent to these terms, prior to the commencement of their participation in the project.[13]

Is the community covenant a well understood mechanism in the communities? Our field interviews suggest that it is not (Mickels-Kokwe *et al.*, forthcoming). This is partly a question of language and contract form. Although covenants are explained in Nyanja, the written text itself is in English, and moreover replete with the legal language of carbon projects. How well do written contracts work in an environment where many of the signatories are illiterate? What does it mean to local people, for instance, that an eco-charcoal project participant commits to 'not engage in, or support, any additional unsustainable charcoal production taking place on his/her own time or own property'?[14] For the village headman, signing a covenant, committing not only himself, but also the community members in his area of jurisdiction, the question is even more poignant. Whom, and what, do the headmen actually commit the community to, when forested areas are set aside for project use?

It is important to bear in mind that the community covenant is not a document signed between two parties, both of whom are linked to other commitments that affect their ability to uphold the covenant. For example, BCP is party to the Carbon Rights Covenant, which is a legally binding contract – the community is not. The village headman, on the other hand, is not responsible for only himself and fellow community members, but, through the Chief, he is a custodian of future unborn generations, who by Zambian custom have rights to the land.

The legal status of the community covenant is also not clear – is it a legal document in a Zambian court? BCP has put in place a grievance mechanism which is explained in the Project Design Document (BCP, 2013, pp.101–102), but does not appear clearly in the community covenants. The grievance mechanism, again explained only in English, includes the mention of a third party arbitrator, either an NGO with localized presence, or a local government official. There are no mentions of the Traditional Authority or local courts in seeking redress. The omission of the courts may be a clear oversight, considering that at least two other institutions involved in the broader partnership, namely ZAWA and the FD, perform a police function, with a clear mandate to arrest poachers and persons engaged in illegal production or conveyance of timber and charcoal.

The point of our discussion here is not to try to find fault with BCP. Rather, we wish to point to what appears to be an ill-defined and complex grey area in the implementation of carbon projects. Despite the best intent and the spirit of the Cancun REDD+ safeguards, project implementation and its national REDD+ guidance has not yet established procedures and protocols that will enable fair and just, well-understood carbon partnerships to grow.

Incentives for participation – community projects

'Yes, I know BCP. They burn charcoal.
They also promote chicken and beekeeping in the area'.[15]

The deforestation mitigation activities under the LZRP take the form of community projects, 'intended to sustainably improve local livelihoods and provide viable alternatives to activities that are dependent upon deforestation' (BCP, 2013, p.139). The Project Design Document lists 17 potential community projects – five that will boost agricultural production, six that generate income from sustainable use of woodland products, three related to the start-up of small businesses, and one each in employment, education and social services (BCP, 2013, pp.140–143). Almost all have been tried in Zambia before, meaning that lessons around constraints and opportunities are available, if not in the project area, then elsewhere.

Partly, the familiarity may be derived from familiar methods of project identification. This was done in an intensive two-month process of consultation with the communities, resulting in the generation of more than 30 project data sheets (BCP, 2013). The process of consultation appears similar to the methodology of the recently concluded JICA-funded Participatory Village Development (PaViDIA) process for community planning and the identification of bankable community micro-projects. The PaViDIA project operated in Chongwe District between 2004 and 2010, mostly concerned with infrastructure projects (Mickels-Kokwe *et al.*, forthcoming).

BCP added a new element by asking the community members to describe what they were willing to 'contribute towards deforestation mitigation' efforts in exchange for the benefits or support they anticipated receiving from the project (BCP, 2013, p.94). All of the projects were to have a community contribution, however small. The proposed community contributions were then incorporated into the conditions for BCP support and formalized in the community covenants.

Most commonly, the communities would commit to behavioural change in exchange for BCP support – for instance, 'those participating in project [sic] will not participate in further deforestation' (BCP, no date). Sometimes, however, a requirement for a tangible contribution is mentioned, especially when small businesses are considered. This may be infrastructure (e.g. a rent-free building to house the project), land (e.g. abandoned fields), information (plans, quote, statistics), or a group action (establishment of a financial or farming group). Communities who would like to launch a project must then write a formal letter of request to BCP, stating their interest in launching the project and making clear their agreement to protect forests in exchange for BCP support.

During our field visits, community members frequently mentioned one of the key community projects started so far, the improved small livestock production (village chicken) project. In early 2013, the BCP Community Engagement Team launched chicken projects in each zone. Small groups from 23 villages were selected to participate in the pilot project. Covenants were signed, whereby the ten group participants committed themselves to abstain from deforestation and hunting in the Rufunsa Conservancy, to care for and maintain the chickens, and to provide feed, vaccinations and shelter, in exchange for ten chickens, feeders and drinkers, training and supervision from BCP (BCP Trust, 2013).

Community members were sceptical of the village chicken project from the start. At a meeting held in Chilimba Zone in October 2013 they complained of the group approach to chicken rearing, the difficulty of the pass-on system of giving all chickens to one person when the anticipated multiplication of birds did not materialize, and quarrels about contributions towards feed and medication. By January 2014, many of the birds had died, either from disease or being eaten by vultures, and only a few groups were doing well. The District Commissioner pleaded for the distribution of sturdier animals such as goats. In May 2014, it was reported that the members of the chicken groups were shunning meetings and had lost all interest. One interviewee said: 'People want to see better, meaningful income generating activities. Chicken keeping is not good. How can you share ten chickens in a village of over 20 or more households? How can ten chickens stop people from cutting trees?' Another person added: 'the chickens are not doing very well [...] the chickens given to one village are very few. If people, in every village, were given his or her own chicken to keep, probably it would work. Group things are difficult to manage' (Mickels-Kokwe *et al.*, forthcoming).

In summary, the community members were disappointed with the implementation modalities of the chicken rearing project. They disliked the group approach, and pass-on strategy. The basic rationale was questioned – would chicken rearing really stop deforestation? None of the community members objected to the condition to abstain from deforestation per se, but did not see a logical link with chicken-rearing, not least because most beneficiaries were women, who were less involved with deforestation activities. Meanwhile the chicken project encountered the usual problems of poultry rearing in rural Zambia – diseases, predators, husbandry and feed. As an organization, BCP appears poorly equipped to address these kinds of difficulties – which lie well outside its core mandate – relative to more experienced government extension services or rural development NGOs.

Conclusions

The challenges facing the new carbon offset projects in Zambia, exemplified by the LZRP, show that, in many ways, carbon projects are no different from conventional development projects. They cannot escape their socio-cultural and historical context, and they encounter similar problems in effectively mobilizing and engaging communities; building capacity; finding viable and sustainable solutions to community problems; and forging viable long-term partnerships. Carbon projects come on top of, and add a new layer to, long histories of such projects and challenges.

At the same time, due to their different design, carbon projects face a greater challenge in explaining themselves: who they are, what they do and who is behind them. They face particular issues of transparency in revealing the ultimate aim of their engagement – carbon trading for profit – and must be prepared to justify themselves to a broad constituency. The requirements of carbon projects – including the need to implement complex protocols and safeguards, rapidly – run against the kind of slow, patient, painstaking processes that have been shown to be so essential for rural development, in Zambia and elsewhere. The LZRP experience raises even more profound issues of authority, legitimacy and governance than earlier projects, yet without offering sustained means to tackle them.

Finally, the LZRP experience highlights fundamental contradictions between the timeframes of carbon projects and the realities of social change. No project in Zambia's history has had a 30 year time horizon – and even state activities are not geared in this way. Rural Zambia now is not what it was 30 years ago, nor in a changing global and local context will it be the same 30 years from now. Engagement over such a timeframe necessarily requires flexibility, learning and adaptation to respond to ongoing dynamics. Here lies the bottom line for carbon engagement – how to find an adaptive, fair, continuous form of grounded engagement, which contributes to both community development and sustainable common futures for all.

Notes

1 M. Kokwe and P. Matakala, REDD+ consultants, personal communication, 27 October 2014.
2 http://unfccc.int/meetings/cancun_nov_2010/items/6005.php, accessed 19–20 October 2014.
3 Unless otherwise referenced, this presentation of the LZRP is based on the official Project Design Document, dated 8 March 2013; a visit to the Rufunsa Conservancy in July 2014; interviews held with BCP Managing Director and the community engagement team in August 2014; a written contribution by the BCP Trust Executive Director in August 2014; and website www.biocarbonpartners.com, accessed 10–11 July and 19–20 October 2014. Our field observations suggest that project implementation arrangements are being adapted and modified continuously. Hence, official project document and website data do not always correspond to realities on the ground. Whilst we have to our best ability tried to accommodate the changes, the authors wish to emphasize that some factual observations may reflect a past development rather than a current project stage.
4 The BCP LZRP initiative is commonly referred to by the locals as 'Ba Carbon', 'the carbon people'. This was first recorded at the Chilimba Community Meeting, December 2013.
5 See www.biocarbonpartners.com/lowerzambeziredd-project/, accessed 19–20 October 2014.
6 See www.biocarbonpartners.com/lowerzambeziredd-project/, accessed 19–20 October 2014.
7 See www.biocarbonpartners.com, accessed 19–20 October 2014.
8 See www.biocarbonpartners.com, accessed 19–20 October 2014.
9 Anna Toness, Economic Growth Team Leader, USAID, personal communication, 11 April 2014.
10 See www.biocarbonpartners.com/our-partners/, accessed 19-20 October 2014.
11 Unless otherwise referenced, this section is based on focus group discussions and key informant interviews in the four community zones – Chilimba, Namanongo, Ndubulula and Mweeshang'ombe over the period February 2013 to July 2014. We interviewed members of Zone Development Committees, Community Mobilizers, village headmen, the agriculture and forestry extension staff, the ward counsellor, members of the eco-charcoal and chicken projects, and ordinary members of the village community. We visited Rufunsa and Chongwe districts, met one of the two District Commissioners, and several members of staff at Zambia Wildlife Authority and the Departments of Forestry and Agriculture. We have deliberately granted our interviewees anonymity. However, lists of persons met and interview recordings are found in Mickels-Kokwe et al. (forthcoming).
12 Interview, village headman, Namanongo zone, 11 July 2014. See Mickels-Kokwe et al. (forthcoming).
13 BioCarbonPartners Trust. Community Covenant Agreement between Sustainable eco-charcoal project participant and BCP Trust. Sample not signed, obtained from member of charcoal group, Ndubulula zone, 11 July 2014.
14 Ibid.
15 Interview with village headman, Mweeshangombe zone, 11 July 2014. See Mickels-Kokwe et al. (forthcoming).

8

STRUGGLES OVER CARBON IN THE ZAMBEZI VALLEY

The case of Kariba REDD in Hurungwe, Zimbabwe

Vupenyu Dzingirai and Lindiwe Mangwanya

Introduction

Experiments to deal with climate change are presently unfolding in Zimbabwe. The Kariba Carbon REDD Project (KCRP) is one such experiment. In order to explore whether these initiatives are successful in their dual aims of reduced deforestation and poverty, we need to trace backwards and see how such experiments are constructed and presented, and ask who ultimately wins and who loses out. For such experiments come on the back of a long history of related environmental interventions, where questions of equity and disenfranchisement have been raised. These concerns are at the heart of this chapter.

This chapter, therefore, has two purposes. Firstly, it is about how these experiments are framed by those both advocating them and desiring their swift implementation. This is a reborn and allegedly 'people-led' private sector that increasingly sees opportunities in the environment, particularly in customary lands, reimagined as vast carbon sinks. Until the 1980s, the private sector did not assign any value to these customary lands, save as a pool for African labour necessary for Zimbabwe's settler economy. The second concern seeks to understand how such carbon experiments are perceived by local groups, and why. Indigenous groups, migrants or women may have different thoughts and perceptions about a project. In this chapter, a socially differentiated set of perceptions are therefore documented and linked to diverse modes of livelihood in the Zambezi valley.

The KCRP has been established in the Zambezi valley, along the shores of Lake Kariba, and stretching over 1.4 million hectares (ha). The project will increase coverage to two million hectares in the next five years. It promises to create value from protecting forests and so generate up to 51 million carbon credits. It aims to share the benefits over 30 years from selling carbon credits, with 30 per cent going to the community, 30 per cent to the Rural District Council (RDC), 10 per cent to safari companies leasing land and 30 per cent to the carbon company. This ambitious project is fast unfolding, and links to the wider policy debate about climate change in Zimbabwe, a country that has experienced climate change and is finding its feet in terms of response (Government of Zimbabwe, 2013). Data for the chapter comes from reviews of project documents, interviews with proponents and participatory workshops with some communities in Hurungwe District, whose Chundu ward is a participant district in the project.

In Zimbabwe, climate change policy has provided an opportunity for land grabs or 'new enclosures' (White *et al.*, 2012), particularly by those white settlers now barred by the neo-nationalist state from owning land as of colonial right. But the land grabs – or green grabs as they are sometimes labelled (Fairhead *et al.*, 2012) – are presented as a missionary venture to save people and their ecology. Yet, in practice, the project undermines livelihoods, forbidding access to foraging, agriculture and hunting across large areas that were traditionally used. All this makes the KCRP strongly opposed by migrants – mainly Karanga – who need land for petty commodity production, and women and hunters who forage and hunt in the forests for household food security. The KCRP is also opposed by squatters wanting land and who wish to avoid intervention that would entangle them with the state and the private sector.

But while the project is a route to dispossession and disempowerment, the KCRP has also become a powerful tool for others. This is a society punctuated by migration and ethnic strife, and indigenous people – the Korekore – who have for years detested squatters and migrants for taking their land, are finding in KCRP a convenient tool to control or prevent further land invasions from outsiders. Rather than a simple, overriding hegemonic force, the image of the carbon project as a focus for local power plays – centred on the control over land – with diverse winners and losers therefore emerges from the study.

Before turning to the details of the case, we now turn to an examination of the wider policy context in Zimbabwe, and the origins of such experiments.

Zimbabwe, climate change and the KCRP

For several decades now, and with support from donors, Zimbabwe has been seized with developing climate change policies. The engagement with this process has intensified in the last few years. In 1992 the government participated in the famous United Nations Framework Convention on Climate Change (UNFCCC). In 2009, the government followed by lending its support to the Kyoto Protocol, and has continued to support a range of UN initiatives aimed at controlling processes contributing to climate change. The government presented a policy document – 'Zimbabwe National Environmental Policy and Strategies'[1] – and set up a climate change coordinating office within the Ministry of Environment and Natural Resources. The policy, together with the Mid Term Plan 2012–2015, became the basis of a consolidated draft document: 'Zimbabwe National Climate Change Response Strategy' (2013). The process ran parallel to the country's initiatives with the regional body, the Southern Africa Development Community (SADC), chief of which was support in 2011 to the SADC-REDD Programme,[2] designed to assist member countries to respond better to climate change and its impacts.

This energetic engagement with climate change and deforestation issues stemmed from the observation that the country's agriculture, the mainstay of the economy, would be affected by increased temperatures and ensuing drought, hitting hard on the 70 per cent of the country's population living in rural areas. Other sectors would be affected, albeit not to the same degree. The result, according to ex-Minister of Environment, Francis Nhema, would not just be lack of development, but also insecurity:

> Climate change threatens water and food security and the allocation of resources which could in turn [...] raise tensions and trigger conflict [...] It is my belief that our sovereignty as a country is compromised if these environment challenges [...] are not addressed.
>
> Herald, *26 March 2013.*

In another interview the former Minister remarked:

> What we are saying is that these effects of climate change are real and we might have to change the way we do things in agriculture, industry and water resources management. We cannot plan according to the old seasons anymore because they are fluctuating.
>
> Xinhua News, *6 June 2013.*

With leadership from NGOs also seeking to tap donor support, a variety of projects and plans have been formulated. These have included land use changes, investments in water conserving agriculture, new energy technologies and so on. Meanwhile, the private sector has stepped into the fray, offering commercially backed alternatives under the umbrella of the government's nascent REDD programme. The KCRP has been the first and most prominent. In its documents it claims it will simultaneously address deforestation, alleviate poverty through forestry-related development and sequester carbon in large amounts to be sold on international markets as part of international offset schemes. A win, win, win scenario – or successful model – is presented, addressing both local and global challenges.[3]

The intention of the founders of the carbon project is to pilot this project in the Zambezi valley, covering Binga, Mbire, Nyaminyami and Hurungwe. This is a giant project, covering 5 per cent of the country's land area of 39 million hectares. In the section below, Hurungwe District, the case study area, is briefly described, focusing on its history of development, its people and other factors that have precipitated the emergence of the KCRP.

Hurungwe: Development and deception

Hurungwe is a district in the north of the country, with a population of 342,675, an area of 19,843 square kilometres and so a population density of 16.6 inhabitants per square kilometre (Zimbabwe National Statistics Agency, 2012). With rainfall averaging 700–800mm per year (Chimhowu and Hulme, 2006; Silber and von Laer, 2012), the area is semi-arid. Large reserves of forests and woodland – about 13,140 hectares – gives the area a huge potential for development. Wildlife is resident in the communal as well as the large and multiple protected areas on the edges of the district (Bird and Metcalfe, 1996; Metcalfe, 2003). For example, in the north is the giant Mana Pools National Park, covering 2,196 square kilometres. To the north east is the Sapi Safari Area, which is a little over 1,180 square kilometres. In the east is the Chewore Safari Area, which is approximately 3,000 square kilometres. It is this remaining forest land bounded to the north by the majestic Zambezi River that the carbon project aims to save by creating new value through carbon.

This region has inspired much romantic reflection on the wonders of the African 'bush', and the Zambezi River and Lake Kariba are characterized both as a wonder of nature but also the result of human mastery over it. These imaginations are held by different people in different ways, both by whites who see this as one of the last great wildernesses in Africa (Hughes, 2006a, b) and by local people who see the river and its banks as a source of livelihood and spiritual control (McGregor, 2009). For others, the valley is seen as a frontier, a place to be tamed, conquered, settled and exploited, and numerous projects involving clearing tsetse fly and encouraging cotton production have been witnessed in the area (Worby, 1995). Equally, new settlers from the crowded communal lands to the south have also seen the area as a place of plenty, and a source of new opportunity (Chimhowu, 2003). It is these conflicting imaginaries of landscape and its value that come together in the contested politics of the KCRP.

Hurungwe District is therefore a melting pot of different peoples, each of whom has come to live and work there over time. The first group are the long-term residents, the Korekore, with a mythical origin north of Zambezi. Tracing ancestral descent from the tribal founder, Chimombe, the group claims a long stint on the Zambezi River where they fished, farmed and kept small livestock.[4] United by a common cult symbolized by spirit mediums without borders and frequently dividing time between villages and sacred forests (Spiereneburg, 2004), the Korekore are now located in villages created in colonial times that include Chitindiva, Buthamocho, Kabidza, Chisauka and Chundu, separated by strips of forests. For subsistence, the Korekore mainly grow traditional crops, hunt and forage, the latter dominated by women and children. Remittances, both national and regional, are also key, with the diaspora using these to register claims to land and forests simultaneously eyed by developers. While the situation is changing due to the growth of lucrative tobacco growing, the Korekore remain comparatively poor, providing the carbon advocates and other donors an appetite to intervene in the name of poverty alleviation.

The second group are migrants, mainly Karanga from southern Zimbabwe and recognized cotton and tobacco growers, who are accumulating land and resources rapidly. Migrants were recruited by local chiefs anxious to increase their political following and to raise the Hurungwe population to levels sufficiently high to attract development sympathy from the state. By locating migrant settlements on the frontier, chiefs shielded their people from nature, particularly marauding animals and the devastating tsetse fly (Bird and Metcalfe, 1996). These accumulators are thoroughly religious, belonging to the many independent evangelical Christian churches. Their religious ideas have brought them into collision with long-term residents, especially their disregard of sacred rules relating to the fertility of Hurungwe. In pursuit of wealth, migrants do not spare sacred forests or wetlands, for example.

The third group consists of 'squatters' or refugees. After the land reform following 2000, many former farm workers from the Karoi farms that were invaded to the south unilaterally settled themselves. These were what James Scott (2009, pp.40–41) refers to as 'non-state places' – remote sites beyond the reach of the state and beyond even the peripheral Karanga settlements. The intention of these latest migrants was to conceal themselves in these thick forests from the vision of a state that was not providing them protection or support (Rutherford, 2002). Occasionally, these squatters got endorsement from a supportive leadership, both bureaucratic and political: for example, migrants who settled in the buffer zone designed for commercial hunting pointed to a former chief as their patron. This has resulted in a local political tussle, with the former chief being accused of hiding former farm workers, and so for flirting with an opposition party supported by the British, and for taking up forest land destined for investment, including by the Chinese.

A fourth group are white hunters and game enthusiasts. In the 1980s and 1990s, white Zimbabweans invested heavily in the hunting enterprises in the Zambezi valley. These were lucrative opportunities given the huge game populations in the area and the growing tourism sector, with tourists from across the world wanting a taste of 'wild Africa'. With the advent of the CAMPFIRE programme (Bird and Metcalfe, 1996), the scope for hunting expanded beyond the allocated hunting blocks and safari areas, and tourism also grew with elite lodges established outside the national park areas. This small group of individuals established a number of well-known companies, including Mburungwe Safaris[5] in Hurungwe, and were well integrated with the Rural District Council, chiefs in CAMPFIRE areas, national parks and other government offices. With the changes in political atmosphere in the 2000s, white entrepreneurs had to change tactics, especially as the tourism and hunting businesses collapsed. This is when the carbon business became an alternative to be explored.

Even before the planned development of the carbon project, local processes and politics were already influencing the forests in Hurungwe. In terms of development, the district has been saturated with planned developments, each with a carbon legacy. Early colonial land grabs scattered the Korekore into the 'reserves' to create

Karoi farms, although justified in terms of protecting their tribal identity. The Korekore who had been displaced in the Zambezi valley as fishermen and crop farmers were to be displaced again in the 1950s to spare them from inundation caused by a huge federal engineering project, the Lake Kariba Hydro-Electric Power Station (Chimhowu, 2002). Immense forest clearance to enable agriculture followed this forced settlement. In the early 1960s, as part of the government programme to create protected areas, the Korekore were again removed from the Mana Pools Park, stretching up to the present day Rekomichi Tsetse Research Station. In the years to follow, a government tsetse control programme would shuffle them further to other forested places up the escarpment (Chimhowu and Hulme, 2006). The 1970s war of liberation orchestrated from Zambia further disrupted the district and indeed the entire valley (Lan, 1985). The Rhodesian state, supported by South Africa, constituted protected villages in Chitindiva to minimise interactions of guerrillas with the local people. A village head in Kabidza remembered that:

> The settler government called everyone to come back from the forests. We had found a good place in Kabidza when we ran from inundation. The government announced in the 1970s that everyone was to pull back to settled villages for protection from terrorists.
>
> *Kabidza, village head, Hurungwe, 3 March 2013.*

As a result of this compressed settlement and a ban on forest-based settlements, massive vegetation changes took place. This continued a decade after independence in the 1990s, when a massive tsetse eradication campaign funded by the European Union followed, preventing movement of local people and livestock for a time. The programme was followed by state-sanctioned rural resettlement, and enterprising migrants from the southern part of the country were encouraged to settle in the area beyond Chitindiva, and specialize in cotton production (Chimhowu, 2002). Cash cropping was accompanied by massive clearance of forests, peaking in the late 1990s (Murwira, 2003), when a flurry of energy and tree-planting projects were initiated to save the environment, yet were later abandoned due to cultural insensitivities.

From the late 1980s, the CAMPFIRE programme took off. This was a wildlife programme premised on 'benefit sharing' and 'sustainable utilization', with benefits shared between hunters and local people via rural councils (Murphree, 1993; Murombedzi, 1992). CAMPFIRE divided the district into three sport hunting concessions (Bird and Metcalfe, 1996), conservation of forests and now valuable wildlife was encouraged. However, many challenges emerged, including the fact that benefits were relatively small and not widely shared, benefiting a local elite and the council more than the wider population (Dzingirai, 2003). The programme completely collapsed in the 2000s, with the crash of the economy and the decline in hunting revenues. Accusations of corruption rose, and the revenues dwindled. Everywhere in Zimbabwe where CAMPFIRE was implemented (Dzingirai *et al.*, 2013), this was followed by multiple strategies to eliminate wildlife. When the

KCRP arrived, offering a benefit sharing scheme based on the use of forests and in collaboration with the RDC, they strategically distanced themselves from CAMPFIRE due to its recent loss of reputation.

This disruptive history of landscape intervention has a legacy, both on the environment but also on people. This has had a huge impact on the KCRP. Where forests are, and who controls them, is heavily influenced by this long and complex settlement history. The earlier bans on settlement and the expansion of protected areas, for example, have meant that the district can boast large stocks of carbon. Yet the availability of land means that it attracts the attention of outsiders, and those desiring refuge of some sort. This establishes conflicts between new in-migrants and long-term residents, and those eyeing the forests as a commercial opportunity in the carbon market. Finally, as a marginal area outside the orbit of state patronage, the lack of development means that people remain poor, with few schools and clinics. The allure of development remains strong, and this is why carbon-based development with its promises of provision remains attractive. Above all, this history of social, ecological and landscape change provides the basis for local understandings and political responses to the KCRP today, and is a lens through which the particular challenges of the carbon project must be seen.

The Kariba Carbon Project

As indicated earlier, the KCRP project covers about 1.2 million hectares of Matabeleland North, Mashonaland West and Mashonaland Central provinces, with their large stocks of trees. Midlands province is involved as a reference area, and so will not get benefits from carbon. The project's rationale centres on reducing carbon emissions through the minimization of deforestation and degradation in those areas, and the selling of carbon credits through international offset schemes. The project plans to reduce deforestation by 10 per cent each year for the first seven years, and then maintain a 70 per cent level reduction for the full project period of 30 years. To run from October 2011 until September 2041, KCRP has achieved CCBA double gold validation from the Climate, Community and Biodiversity Alliance. Further, the project has achieved Verified Carbon Standard. Finally, the project received approval by the National Carbon Offset Standard, becoming the sole project to be endorsed by Australia for use as voluntary offsets by its businesses.[6] All these validation processes were protracted – taking over four years – and involved mountains of paperwork, assessments, inspections, audits and field measurements. As the project developers confirm, it is an enormously complex process. It is also hugely costly, because professionals have to be hired to fill the forms, field teams deployed and expert assessors involved. And all of this before any carbon finance is available. From inception, due to the technical complexities, local communities are already disempowered and are unable to give direction to the project they will live with for over 30 years.

The project is being implemented by Carbon Green Investments Africa, a Guernsey-registered company.[7] The Harare company involves mainly white staff who share a background in safari hunting and large-scale commercial farming. Due to the complex land reform, many were forced off their farms, and hunting also declined due to political insecurity (Scoones *et al.*, 2011). At the time of writing, the company had one senior black employee with a background in the police, and who is in charge of community liaison. The safari background, the company maintains, is a plus to the KCRP for two reasons. First, it provides key lessons about how to deal with and manage the communities and the local council, and, secondly, the safari and CAMPFIRE background also provides environmental knowledge necessary for KCRP execution. As one member of the team explained:

> [T]hese people […] have worked on the ground in these areas for years, as safari hunters and so on. They really know the area and environment.
>
> *Project staff, Harare, 13 December 2012.*

As shall be discussed, this background to the company – and perhaps especially its racial composition – has generated all sorts of suspicion and theories from local people about what the project will eventually do to them.

Finance for the project's start-up has come from venture capital from a local white Zimbabwean entrepreneur, who is deeply committed to environmental issues in the country and has provided around US$500,000. The development of the project is also supported by South Pole Group,[8] together with individual investors. Between them, these investors raised US$750,000, which has supplemented the original US$500,000 from the private investor. South Pole has in excess of 90 carbon professionals, is based in Zurich and has global offices – including in Johannesburg and Kampala. The project, therefore, has good allies and a solid financial backing. It is firmly geared to the commercial arena, and unlike the international donor, NGO and charitable funding provided to previous interventions, concerns with financial viability loom large. As an officer of the company informally remarked: 'We are in business first and foremost' (Mac, project staff, Harare, 13 September 2013).

In this venture, other actors are involved. First is Environment Africa, a forward-looking regional NGO created by two Zimbabwe women with a passion for ecojustice. The company tries to balance environment and development needs and has won numerous awards in sustainable utilization of natural resources. As a partner in Kariba REDD, the NGO has been tasked with socio-economic feasibility studies. Its work has provided the basis for the strong narrative presented in the project design document that deforestation is caused by villagers' destructive practices:

> through their clearance of land for agriculture, trees for fuelwood or household use, fuelwood for tobacco curing, veldt fires caused by poachers and trees for poles used in the construction of homes and garden fencing.
>
> *Environment Africa, 2011, p.5.*

While questions have been raised about the local deal with communities, the NGO has provided a key part of the justification, based on its 'participatory appraisals' in the field.

The other partner is Hurungwe Rural District Council, along with three more local authorities, namely Mbire, Nyaminyami and Binga. The councils bring no other equity, other than the land they preside over and the requirement to control people from subverting the project through unsustainable uses. Even if it wanted, the local authority has no money, and has no sources other than from levies. Indeed the local authority is more like a sleeping partner, with very little knowledge of what the project is all about and with no voice on its direction. As one of the officers at the local authority remarked:

> These guys have the money, and what do we have? Nothing […] and these guys have the lawyers. They even have experts who know what carbon credit is. Some of us do not know what it is, nor do we know the details.
>
> Hurungwe Rural District Council Officer,
> Hurungwe, 24 November 2013.

Despite being the formal 'owner' of the land on behalf of the state, the council has little influence. Instead, much of the running is made by safari companies who have hunting leases across the area issued by the council. Mburungwe Safaris are partners in the project and land leaseholders in the area, and so will receive 10 per cent of the share of proposed carbon revenue. This applies only in Hurungwe where these brokers are found. Yet there is tension between the different players, and the safari companies are not always regarded favourably by the local communities or the council, as they have been accused of not presenting the full accounts and sharing revenue fairly under CAMPFIRE agreements. Nevertheless, Mburungwe Safaris has faith in the council and has renewed its lease with the authority. The company, which runs anti-poaching operations in its concession area, sees the KCRP as providing an opportunity to rejuvenate forests in ways that will support wildlife, as well as providing an additional revenue stream.

Each of these key actors are united in the view that communal resources, including forests and wildlife, are there for the taking, that local communities are not using them effectively, and that new forms of private sector-led management will bring improved management and better environmental protection. This view, however, is not shared by local people, who regard the initiative with scepticism and suspicion.

KCRP and missionary work

The KCRP presents itself as a missionary engagement on two fronts: tackling poverty and improving ecology.

As explained by the project developers, the project is seen as a missionary engagement run from personal sacrifice because 'We have been lucky, (the boss has) invested out of commitment; no banker would touch this, returns being uncertain and over time' (Project staff, Harare, 13 November 2012).

It is its missionary disposition that propels the company to assist the councils, who it is at pains to argue are the formal implementers and owners of the project on behalf of the local populations. This was a key part of the pitch to achieve registration at CCB Standard, Gold Level. Further, it is the missionary disposition that pushes the company to implement in the project to save local environments and to contribute to the global challenge of mitigating climate change.

Its other business is to banish poverty, and this missionary focus will target the nearly one million poor people in the Zambezi Valley by providing a new source of revenue to local communities from the sale of carbon credits. The design document displays this ambition, with poor children on its cover, choosing for maximum effect barefoot schoolchildren with no uniform, and having a poorly patched grass thatched structure as the classroom (Kariba REDD+ Project, 2012). This imaging of the project as a missionary endeavour is also conveyed to the local people. All the district councils have been provided with artistic impressions of the project beneficiaries. One popular cartoon stuck on council office walls shows poor women pounding grain, while malnourished children wait. And all this in the background of a failed crop, perhaps because of climate change. The cartoon is captioned: 'Why we need KCRP, get more details from the Council'.

An end to deforestation, assured by the reduction by nearly 10 per cent each year from year one to year seven, and then by 70 per cent for the full project period of 30 years, will create what project documents describe as 'good forests' or 'good habitat'. An artist's impression captures the scenario with Kariba REDD. In this impression, land is restored to its Edenic status, with large forests concealing even homesteads. It is these regenerated forests that will produce nearly 51 million credits to be sold in offset schemes, with proceeds to be shared according to a fixed formula. Although the project focuses on trees and carbon, there are many resonances with the CAMPFIRE programme that worked through the council, and shared benefits. But the KCRP plans a different arrangement, however:

> A significant (20 per cent of net profit) share of the project's carbon income will be invested into general activities promoting and guaranteeing sustainability of the project [...] The project is [...] on communal lands and as such it is imperative the people within these communities have their lives enriched by the project
>
> *Kariba REDD+ Project, 2012, p.34.*

We (plan) to give the RDC their share but also keep an allocation for the communities separate. We can understand what is going on, we can manipulate and channel the money.

Project staff, Harare, 13 November 2012.

How this will be done, given that the RDC is the official representative of local people, is ambiguous, and will surely be resisted by the local state. In the absence of policies dealing with REDD projects in the country, only the RDC is empowered by law to enter into agreement about how benefits from the land ought to be used. This means, as with CAMPFIRE, the RDC has ultimate control, despite the rhetoric of community involvement. Further, there are practical challenges of distributing benefits to heterogeneous communities and diverse individuals. The project can expect a political minefield to confront it, as other past experiences have shown. As a Kabidza resident noted:

> It will be the problem we saw in CAMPFIRE. I will be expected to respect trees, and not expand my field to earn more money from cash crops. Yet when the money comes it is shared equally with those that cut trees.
> *Chisauka, village and religious elder, Kabidza, 2 June 2013.*

Clearly the problems of land administration politics and tenure is such that carbon revenues will probably not, as in the past, find their way to the local level in ways that change poverty. KCRP argues that it will be transparent in the manner in which it handles returns from carbon. But what has happened to date does not offer encouragement. Nobody knows, including the RDC, how much has been generated since the KCRP started, if anything. The only funds that have exchanged hands are US$200,000 allocated to the council. An RDC official observed:

> I am telling you we are in the dark about what these guys are doing. These people are not straightforward at all. At least in CAMPFIRE we knew how much we made, but these guys keep us in the dark.
> *Hurungwe Council Officer, Hurungwe, 10 November 2013.*

But it is not only the council that is in the dark about how much money is realized from carbon. Even the well-meaning NGO, Environment Africa, which has been promising people money from trees in its community liaison work, does not know, and this makes them feel guilty each time they are in the community: 'Imagine, we have been working on this thing together and yet we are not told how much has been realized from carbon. To the communities we appear liars, and the feeling is not good' (Maware, Environment Africa Officer, Harare, 11 September 2013).

Without this transparency, it is doubtful that anything can even come from carbon that changes livelihoods, and conflicts are sure to increase.

Livelihood benefits?

In order to persuade people to save the forests and capture carbon, other projects to improve livelihoods are being proposed. These are centred on improving agriculture, and in particular banishing 'slash and burn' practices. The design document makes great play of the ravages of 'slash and burn': 'Subsistence agriculture on plots [...] often based on slash and burn methods, thereby leading to significant deforestation' (Kariba REDD+ Project, 2012, pp.18–19). The KCRP intends to change all this, and promote a more sustainable, conservation-oriented agriculture, focused on small gardens, improved water infrastructure and 'conservation farming'.

At face value this all appears sensible. However, the argument is premised on a number of false assumptions. First, that 'slash and burn' is a dominant agricultural practice in the area. In the past, forms of shifting agriculture certainly existed, but this has not been significant for decades, and today most farming takes place in extensive dryland fields. Second, that a shift to intensive small-scale irrigated farming makes much sense, if the water is available. Experience with borehole drilling programmes has been disappointing, as a local resident explained:

> They fight and fight every day. They fight for the water. And they fight for the turn. One says it's my turn and the other says, my bucket was placed in the queue before you came. And so people fight.
>
> *Chirindo, village head, Mahwai, 5 October 2013.*

Intensive soil and water conservation systems – notably 'conservation agriculture' involving concentrating water and nutrient in pits – have not been popular due to the requirements of labour (Boudron *et al.*, 2011). As one resident explained:

> It's because you must be in the field all year round. You must collect stover after harvest. You must store it somewhere. Then you must make basins, one by one using the hoe. Then you must take stover back to the field, and make mulch. You have no time even to make beer for your ancestors. We will dig the basins and die.
>
> *Chipato, villager, Hurungwe, 8 October 2013.*

And in any case, people point to the abundance of land and the advantages of clearance in reducing wildlife damage and tsetse fly challenge. As explained by one resident:

> If you leave these forests standing, then you have a problem of wildlife. The elephants will not bother you, because they won't have shelter to hide in. And the tsetse fly will also go away. Tree clearing is natural fencing.
>
> *Bocho, villager, Hurungwe, 8 October 2013.*

Livelihood benefits will also be provided through non-agricultural income-generating activities. A whole array is proposed from beekeeping to plantation woodlots. For example, the project envisages communities organized under beekeeping associations, producing honey for their own consumption and also for the market:

> We do not know how this is done, but we do think a couple of look and learn tours to place like Mutoko Beekeeper Association will help us move forward. The founders of KCRP Beekeeping will yield as much as [US]$500.
>
> *Project staff, Harare, 7 August 2013.*

As with so many other small-scale development projects that have been tried in the past, there are many challenges, including skills, markets, organization and the questions of intra-household distribution of income. Given past experience, there is no certainty that any of these many projects will provide broad-based livelihood benefits, and so provide the incentives for people to switch their activities from making use of forest areas, and clearing land for agriculture.

KCRP and wildlife utilization

Building on the experience of the safari operators in the KCRP team, the project also plans to introduce wildlife utilization initiatives in the project. The utilization will target big game like elephants, buffalo and lions. Local people will have the opportunity to enjoy revenue from wildlife, which they will use for local development.

The biggest problem in wildlife management schemes is that they do not provide communities real ownership of wildlife (Hutton *et al.*, 2005). Instead, tenure is accorded to the local authority (the RDC) by the Parks and Wildlife Authority of Zimbabwe. Without this tenure, communities cannot determine how the wildlife ought to be used, by whom and according to what formula, and power is held by the local authority. The danger is that local authorities, including Hurungwe, abuse this privilege, using it to monopolize benefits (Dzingirai, 2003).

If revenues are to be generated from valuable large game, a new form of control and securitization is envisaged for the area. The document explains: 'KCRP will reduce the pressure on the wildlife in the area […] and relieve wildlife from poaching through the project's patrolling [and] increase the number of patrols and number of man days spent patrolling per year, monitored continuously, reported upon verification' (Kariba REDD+ Project, 2012, pp.77–79) One of the fears that women have is that foragers, who depend so much on the forest for tubers, roots, edible worms, fruits and insects, will not be free to continue with their forest-based livelihoods. As one woman remarked:

> Even if they do not shoot us, we will still be afraid of these anti-poaching units, who will be moving with guns in hand. Will we not be confused of being poachers? We will not go to the forest as we have always done.
>
> *Miriam, widow, Hurungwe, 7 August 2013.*

Protecting forests means protecting wildlife, and revenues can be generated from both. But this also results in exclusions, and a recasting of landscape as one that has to be protected by paramilitary force rather than used by local people.

Missionary ecology

The commitment to preserve and protect fragile ecologies is at the centre of the project narrative. A captivating YouTube clip, sponsored by the funders, shows evidence of vegetation elimination as a result of subsistence agriculture. Similarly, a press statement posted online by the company summarized the ecological missionary thrust, with the project's aim being to: 'protect a vast and vital ecosystem at Lake Kariba and its surrounding forests and communities.'[9]

The project design document describes the calamity which necessitates this missionary intervention. Data is used to support the argument. Zimbabwe, it is claimed, loses nearly 2 per cent of its forest cover every year, becoming ninth in the global deforestation rankings.[10] It is, as one informant indicated: 'So serious that it is like having the size of Zaka district treeless every year. You've got to respond and that is what we did' (Forestry Commision officer, Kariba, 5 March 2013). Another informant agreed: 'There is a disaster that can't be ignored! We are stepping up to this. There is rampant deforestation in the area' (Natural Resources Officer, Hurungwe, 10 March 2013).

KCRP does not just stop at characterizing the disaster, but attempts to find the real cause, namely poor crop and livestock agriculture, and in particular 'slash and burn'. According to this narrative, farmers target virgin land, as the project elaborates online:

> Due to poor resources and knowledge, communities are cutting down vast tracts of forest in order to clear fields to plant their staple crops such as maize, sorghum etc. This agriculture is both on a subsistence and commercial level (small commercial level). The trend within such communities is to move from one field to the next at very regular intervals (1–2 years) subsequently cutting large areas of forest down for the new fields. Our feasibility study showed areas that are demarcated for this year's clearing and it became obvious the area of ground that can be lost to this practice and the speed in which this can happen.[11]

The result is a train of deforestation and according to an implementing officer, 'One that you can't miss as you pass through the project area' (Project officer, Harare, 5 March 2013).

But, as noted already, this is a false diagnosis. Indeed, the field surveys that preceded the project did not observe the practice (Environment Africa, 2011). Instead, what is visible in Hurungwe is a growing mechanization of agriculture and the expansion of tobacco fields. This presents a different environmental challenge through the demand for fuelwood for tobacco curing and the expansion of land for the lucrative tobacco crop (Chivuraise, 2013).

Livestock farming is also condemned. The project claims that smallholders are overstocking. This it says is leading to overgrazing in some parts of the area where goats, cows and donkeys were kept (Kariba REDD+ Project, 2012, p.20). Yet, even in the most settled part, livestock is still very limited due to past tsetse challenges.

Although KCRP sees itself as a response to an ecological crisis caused by poor local practices, the diagnosis it brings and the assumptions it uses can be challenged. This mismatch between an 'ecological missionary' understanding of environmental change and the reality on the ground is sowing the seeds for conflicts and misunderstandings. The next section looks at what villagers think about the project.

Local perceptions of KCRP

The project is interpreted and responded to by different groups of people, depending on their relationship to land and resources, and their priority livelihood strategies. A highly variegated response, with diverse perceptions, is seen.

Long-term 'indigenous' residents

Local Korekore people regard the recent in-migration of outsiders as the main cause of environmental change, including widespread land clearance and deforestation. Alec Miroro observed:

> It's all deforestation [...] Trees are cut. They cut them down every year. Trees are especially cut in the dry season [...] If it continues like this, the result will be desert. Cutting down of trees must be controlled.
>
> *Miroro, villager, Hurungwe, 7 July 2013.*

The same sentiment was shared by Richard Made, who remarked:

> Erosion has taken place to an extent and tree population has decreased. There is deforestation in every village.
>
> *Made, war veteran, Hurungwe, 7 July 2013.*

In a participatory mapping exercise, long-term residents jostled to mark areas that have been rid of forests. A litany of destruction was recounted, including the disappearance of wild fruit trees, wildlife and other resources. All one sees, women noted, are clear and open lands with scars of erosion. Men concurred, one adding that 'Hurungwe is Sahara, because of deforestation' (Marongwe, beekeeper, Hurungwe, 7 July 2013).

This depiction of environmental destruction and impending catastrophe is similar in many ways to the narratives promoted by the project to justify its intervention. However, the speech is not random but carefully selected by speakers to mobilize urgent political action, notably to stop in-migration of 'outsiders'. It is this political context, and the competition for resources between groups, that the quotes reflect, more than a formal scientific assessment of environmental change. These local narratives, just as the wider project narrative, serve, above all, a political purpose.

In terms of causation, the long-term residents are in agreement that the culprit is a new type of agriculture being practised:

> This is not ordinary agriculture which focuses on maize and other small grain. It's not the agriculture we are used to.
>
> *Dandawa, village head, Hurungwe, 6 June 2013.*

Village head, W. C. Chundu, described this as:

> Commercial crop cultivation and especially tobacco. This crop requires large tracts of farmland and more importantly, fuel wood for its curing.
>
> *Chundu, village head, Hurungwe, 7 July 2013.*

Informants argue that this massive environmental change started after 2000 when farmers became obsessed with tobacco.

In terms of KCRP impacts, long-term residents perceive it more as a developmental project rather than as a vehicle for stabilizing climate. However, the nature of this development is disputed. Some mention beehives as the most prominent intervention:

> From my field I get maize, and other food crops. Now KCRP has come in handy. From the beehives they gave us, we now sell honey and have some cash every month. The problem is scale, and once we have more beehives and markets we can sell more and make good money.
>
> *Mavhara, village head, Hurungwe, 7 November 2013.*

Again, those residents who mentioned windfalls from KCRP when pressed rarely produced evidence of real change. Much was still locked up in promises for the future from the project.

However, for local indigenes, the project is less about these development benefits – these are like projects they have seen before – but more about the prospect of the project being a route through which migration will be curbed. There is a strong defence of KCRP amongst this group. In a focus group discussion, two women were visibly angry when KCRP was equated with CAMPFIRE. Forced to show the difference between the two, long-term residents singled out KCRP's emphasis on environmental education. As Chidamoyo noted:

> There is a difference [...] because CAMPFIRE never educates people on resources. We are saturated with knowledge of the environment, especially trees.
>
> *Chidamoyo, teacher, Hurungwe 10 March 2013.*

However, when informants were asked to describe what they learned, they could not remember anything, and confusion reigned when the subject of 'carbon' was brought up. One of the informants said he knew that KCRP 'is about clouds'.[12] Another said it is about 'smoke in the air',[13] while a third said it was 'about our trees'.[14]

The Korekore residents cling to the hope and promise of the project. Project supporters are mainly the original Korekore, who are yet to engage in cash cropping to the degree that migrants do. Occasionally they grow tobacco, and they do so on very little patches of land; otherwise they grow maize and cotton on small patches of land that they increasingly fight for. These residents cannot expand outwards to the north because migrants have taken that land too. This generates bitterness, as the remark from Mr Chisauka shows:

> We are trapped. We have no inheritance, because certain people stole it.
> *Chisauka, village head, Hurungwe, 10 February 2013.*

In this instance, then, the REDD carbon project emerges as an unintended but appropriate tool for dealing with migrants, rather than a tool of development or climate mitigation.

Migrants

The migrants offer a very different view of KCRP. They do not dispute that there is environmental change, and that trees are disappearing over time. Their point is that where trees have disappeared it is because of natural causes, mostly wildlife damage and drought, not the hand of humans. One prominent farmer originally from Manicaland, Mlambo, made this point well during a group discussion:

> I do not understand why people say this is the case. I think this is nonsense which even a grade 7 pupil won't accept. If you can go into Mana Pools you will find that there are no trees there. If you go to Hwange Main Camp where there is not even a single village, you will see deforestation. Are you telling me that there are people or farmers who bring these trees down?
> *Mlambo, Zimbabwe Farmers Union member, Hurungwe, 8 August 2013.*

Another resident, Ramson Shumba from Zimuto, Masvingo, noted:

> [T]rees are drying up; this has nothing to do with tobacco farming. Even if you go to other areas you will find the same, trees drying up. Even in Harare trees are drying up and this is a fact; it is not because of tobacco.
> *Ramson Shumba, tobacco farmer, Hurungwe, 6 August 2013.*

In participatory ranking exercises, migrants identified drought and wildlife damage as the two major influences on environmental change.

As regards to KCRP's promise to deliver benefits, these residents immediately scoff at the idea. With visible anger, migrants say it is another CAMPFIRE programme incapable of delivery. A resident originally from Buhera pointed out why these two were similar:

> They are the same; you can't see any difference between the two. We have benefitted nothing from the two.
>
> *Shava, tobacco farmer, Hurungwe, 8 August 2013.*

One of the oldest migrants in the area, Muridzo from Masvingo, remarked that KCRP equalled CAMPFIRE on account of its empty promise matched by disappointing results:

> [T]here is no difference because their promises are the same. They promise money, jobs and everything, but they never fulfil these promises, only benefit the council and the safaris […] a handful of local elites can also benefit but it's rare.
>
> *Muridzo, tobacco farmer, Hurungwe, 8 August 2013.*

But perhaps the most dramatic reflection comes from Lioness Made, a leading member of a Pentecostal church and a tobacco farmer who came to Chundu after initial settlement in Magunje. Made agreed that KCRP resembled CAMPFIRE because it is biased in favour of nature, not people:

> [T]he difference is very big but in the end they are the same […] they are the same […] because they were started by white people – the safari operators – and their friends at the district councils and others from Harare. Let me say this: CAMPFIRE is about sparing wildlife. This new project on carbon is about sparing trees. None of the two is about sparing us villagers.
>
> *Made, tobacco farmer, Hurungwe, 6 August 2013.*

What is clear from the above quotations is that, according to migrants, CAMPFIRE is associated with everything bad. It is accused of favouring nature; it is accused of being extractive; it is said to be selective in its benefits and it is accused of making empty promises. Equating KCRP to CAMPFIRE seems to be a convenient way of dismissing the project. But why do they perceive it as such?

Part of the answer lies in the identity of the migrants. As pointed out earlier, migrants are not Korekore people. They came to Hurungwe to pursue farming, seeing in it a viable livelihood strategy even surpassing formal employment. As modernists, they should easily understand the linkage between agriculture and ecological change. That they deny tobacco to be a cause of deforestation, and that they claim no benefits from KCRP seems a convenient way of protecting their livelihoods in Hurungwe: they should be allowed to stay, because they are not doing any harm to carbon. By dismissing the project, they are attempting to reassert their claims on the land, and their increasingly lucrative, if environmentally destructive, farming livelihood.

Squatters

While migrants think there is deforestation linked to natural forces, and while long-term residents think that environmental change is due to migrants, the squatters are convinced that all is well. Whether women or men, they point to evidence of this stability. Alec Phiri, a former farm worker from Karoi, expressed it thus:

> Look there are trees standing, there is green grass, and tourists like you are coming to see this beautiful place. Would you come to a desert?
>
> *Phiri, squatter, Hurungwe, 6 August 2013.*

To back their claims that forests are still intact, they point to the RDC and safari operators who are attracted to the land. They say that no one can invest money in a landscape that is rundown, and that the RDC and KRCP are investing money in the erecting of fences and camps means that the land still has its integrity. As one squatter of Malawian descent commented:

> Who will put his money where there is no hope?
>
> *Lumumba, squatter, Hurungwe, 12 September 2013.*

As regards the suggestion that deforestation is taking place and that they are responsible, squatters do not agree. Squatters say that, wherever they are settled, they are living peacefully in the environment and not starting any fires. They say that their area is not characterized by erosion, which may well be true since they are not engaged in cash cropping of any kind and since they do not have cattle on account of their recent arrival into the area. As regards their thoughts about KCRP and its promise, they say it is an unnecessary project, born of superfluity. Mr Chiuta, who is a squatter, originally from Mozambique said:

> It's just you can't tell someone what to do with their money. Otherwise this money, if ever these white people and the government have it, could be better used if it was given to us to as pension for working in the farms. We could build proper houses and provide for our families.
>
> *Chiuta, squatter, Hurungwe, 2013.*

Although condemned by the project, and urged to move, squatters are not confrontational. Their strategy has taken the form of avoiding direct contact with the council and KCRP, scurrying for cover as soon as they see or hear the sound of a motorized vehicle. This has made it virtually impossible for the local authority to serve eviction notices to them. Squatters would rather they were left alone and KCRP never came in the first place. Part of the reason is that they do not see the project's necessity or how any environmental change relates to their action. But there is another reason. As pointed out earlier, the squatters are former farm workers who had running battles with the state on account of their support of white settlers. Vulnerable in the invaded farms, these people fled to Hurungwe's forests where they hid themselves from the state. That they would rather KCRP leave them alone does not mean a hostility towards all forms of modernization and development; rather they seek to minimise development interventions – and carbon is one such – that risks bringing the state, and its violence, close to them again. As Mr Banda, of Malawian descent, noted:

> People should have mercy on us and leave us alone; we are old and need no more than shelter from the trees.
>
> *Banda, squatter, Hurungwe, 9 August 2013.*

Different groups therefore see environmental change and the role of the KCRP in radically different ways. This is less to do with the project and its activities, and certainly nothing to do with global climate change, but more to do with the political position of the different groups, their livelihood strategies and aspirations and their association with the land. It is this politics that is creating the conflicts seen around the project, and the mismatches observed between project plans and realities on the ground.

Conclusion

Climate change policy – and particularly carbon offsetting through forestry projects – has certainly provided an expanded opportunity for some powerful social groups, and the formation of new alliances, in rural Hurungwe. Where tenure is ambiguous and authority dispersed, the potential for new players – in this case an alliance of external private sector investors, safari operators and the local council in the form of a carbon forestry project – to enter the scene expands. While the project justification has been presented in missionary terms – around environmental salvation and poverty reduction – the process has been driven by the enticement of revenues from previously untapped carbon resources.

For some this may seem like a simple 'green grab', where land and resources are expropriated. But there is not a simple monopoly of capture, and the negotiations of access to new resources are complex and creative. Competing narratives and justifications abound. KCRP cleverly presents itself and its experiment as a missionary project responding to potential calamities, social and ecological, local and global. KCRP casts itself as a project of poverty alleviation, and at the same time claims to exist to save biodiversity through carbon. But in order to achieve its aims it must create villains and saviours. Local people are cast as the problem, and through a suite of interventions, livelihood activities are criminalized. The project comes to the rescue through providing alternatives, as new revenues are provided.

But the claims of the project are not universally well received. The project is seen as a threat to land access, and to the fruits of the forest in terms of foraging and hunting livelihoods. In the process, local people can no longer access the natural resources they used to enjoy. This recasts the landscape and its use in new ways, and results in resistance and mobilization, as people seek the support of chiefs, political leaders and others to undermine the project.

Yet there are not only detractors. Some groups see advantages to the project, although not in ways envisaged. The project is seen as a means to assert ownership of the land, and associated traditions, and evict squatters and in-migrants. This is why they sometimes wantonly praise KCRP in their daily conversations.

What then can be learned about the politics of carbon forestry projects from this case study? First, in contexts of ambiguous tenure, elites, whether racial or political, can use carbon, as any other valued resource, to appropriate resources from local people. The resources are multiple and include both trees and land, and their products. Equally, local people can use the promise of carbon to control other local groups from taking their resources. The project thus becomes a focal point for the struggle over resources, and so intensely political.

Carbon initiatives are, however, not as totalizing and hegemonic as they are feared to be. To be sure, these initiatives have a potential to expropriate land quietly through the private sector and its partner, the state. They also have the potential of displacing or occluding other forms of use. But people can redeploy the carbon project narratives about environmental decline and resource conservation to influence their own political struggles, with new alliances struck. Carbon projects are replete with politics, and as they play out on the local stage, new struggles become entwined.

Notes

1 See www.ies.ac.zw/downloads/draft%20strategy.pdf, accessed 1 September 2014.
2 See www.sadc.int/themes/meteorology-climate/climate-change-mitigation, accessed 1 September 2014.
3 See www.carbongreenafrica.net/component/content/article/44-home/50-welcome-to-carbon-green.html, accessed 1 September 2014.
4 Chief Chundu, Hurungwe, 12 August 2013.
5 http://thesouthpolegroup.com/.
6 More useful details on these processes are available here: www.coderedd.org/redd-project/carbon-green-investments-kariba-zimbabwe, accessed 1 September 2014.
7 See www.carbongreenafrica.net/about-us-.html, accessed 1 September 2014.
8 http://thesouthpolegroup.com/.
9 www.southpolecarbon.com/_downloads/PR120221_990Kariba.pdf, accessed 1 September 2014.
10 www.southpolecarbon.com/_downloads/PR120221_990Kariba.pdf, accessed 1 September 2014.
11 www.carbongreenafrica.net/home/41-our-projects-/46-kariba-redd-project.html, accessed 1 September 2014.
12 Shumba, beekeeper, Hurungwe, 10 March 2013.
13 Huyo, villager, Hurungwe, 10 March 2013.
14 Chimombe, villager, Hurungwe, 10 February 2013.

9

FARMING CARBON IN GHANA'S TRANSITION ZONE

Rhetoric versus reality

Ishmael Hashmiu

Introduction

Global narratives about the new carbon economy present large ambitions. But is the rhetoric really attuned to local realities? Due to weak or contested land rights, forest carbon interventions can even restrict poor communities' rights and access to forest resources (Springate-Baginski and Wollenberg, 2010; Vhugen *et al.*, 2012). Understanding the situation on the ground, and linking local realities to wider policies, plans and project designs is essential. This is why a detailed analysis of existing projects – even ones that have ostensibly failed – is essential to learn lessons for the future.

This chapter focuses on the Carbon Credit Project (CCP) in Ghana.[1] From 2008, this operated across the country, but was focused in the Transition Zone of the Brong Ahafo Region; predominantly in the farming areas around the towns of Sunyani and Wenchi. Although the data is inadequate, project documents suggest that about 300,000 farmers were engaged throughout the country and 560,000 hectares of land bank was secured. The project design was premised on wealth creation and agroecological sustainability through carbon farming. It managed to gain considerable local attention through the media, but did not respond well to the realities on the ground. The CCP was for many years the only active forest carbon project in Ghana. Although the project was widely dismissed as a poorly thought out private sector venture that was not aligned or attuned with the national REDD Readiness process, the project, nevertheless, provides a good case for examining the tradeoffs between carbon forestry and agriculture, as it was tailored to enhancing carbon stocks and food production in agricultural landscapes. Issues relating to land tenure, resource access and livelihoods, highlighted some of the core dilemmas and challenges of REDD-type projects in agricultural areas, all come to the fore, revealing important lessons.

In the following sections, the chapter explores the environmental and land use change contexts of the project area in the Transition Zone. It draws out the narratives of the key actors about forest change, carbon and climate change and how these narratives converge and conflict. Power relations of various actors and how they influence resource access and control in the CCP are also discussed. The chapter finally analyses the implications of the CCP for agricultural livelihoods and land-use change in the study sites and the Transition Zone more broadly. The chapter brings to the fore alternative narratives from the grassroots that had been obscured by the dominant narratives of the more powerful actors. Such counter narratives from the grassroots potentially provide policy spaces for making carbon offset interventions more equitable, while enhancing the inclusion of marginalized local actors. The design of the CCP itself, the challenges faced by farmers and the project developer, as well as factors that ultimately contributed to the collapse of the project, all provide useful lessons for ongoing and future forest carbon projects worldwide, especially with regard to making them grassroots-centred and compatible with smallholder agriculture.

Environmental and land-use change contexts

Any external intervention, be it a carbon project or not, does not start from a blank slate. It must engage with past histories and experiences and complex socio-ecological contexts. The CCP was no exception. This section outlines some of these contexts, focusing on perspectives of environmental and land-use change in the Transition Zone. The Transition Zone is located between the Guinea Savannah zone to the north and the dry semi-deciduous forest zone to the south. The landscape is a blend of the Northern Savanna and Southern Forest vegetation – hence the name, Transition Zone. The soil and climate are suitable for both savanna and forest crops. Crop farming (mainly maize) is therefore the main source of livelihood and this attracts many migrant farmers from the neighbouring Northern regions.

Settlement and environmental history

Two of the project communities, Badu (in the Wenchi Municipality) and Dumasua (in the Sunyani Metropolis), including their respective suburbs – Tainso and Mantukwa – were chosen for this study because their ancestral histories give a certain indication of the state of past forests and how the land-use pattern has changed over time. The establishment of Badu, for instance, dates back to around 1830 and the ancestral tale holds that the vegetation at the time was originally small patches of grassland in the midst of forests. The open grass fields were ideal for yam cultivation while the forested areas were cleared for cocoa production. The landscape history of both communities is basically a cycle of forests being exploited for shelter, hunting and farming, and farmlands shifting back into forests through a long-term bush fallow system. Thus, local narratives on forest change revolve less around a pattern of secular decline, but more around the idea of cyclical patterns.

Local accounts also emphasize key events, notably the devastating 1983 bushfires that destroyed forests and farmlands, and resulted in considerable reduction in rainfall and agricultural productivity. The mushrooming of invasive weeds following the fire incidence is reported to have resulted in recurrent fires and increased the labour cost of farming. Essential non-timber forest products (NTFPs) like bushmeat, mushrooms and snails have also been disappearing after the devastating fire. Cocoa farms were burnt down and cocoa production, which used to be the main source of wealth, began to decline due to recurrent fire episodes until the land dried and was no longer able to support production.

Farmers therefore believe that reforestation can help rehabilitate the environment, curb wildfires and improve livelihood opportunities. This provides a firm basis for carbon offset interventions. However, any such intervention must occur within a highly dynamic landscape ecology. Justifying additional benefits of carbon sequestration is particularly challenging in the Transition Zone as it is a dynamic system where forest, bush fallow and savanna are in continuous flux.

This presents a challenge for carbon projects that start from the assumption that a secular decline in forest cover is being reversed. In proposing a project, it is, thus, difficult to say that there is a clear baseline, due to highly uncertain fluctuations in carbon gains and losses as forests shift to savanna and back again as a result of complex settlement and agroecological histories. It is indeed this dynamic interaction of fire, fallow and farming that makes the landscape (Amanor, 1994). A baseline forest cover level is in some ways antithetical to a sound ecological understanding of this system. Of course, there have been overall changes within this dynamic, but the technical requirements of the carbon offset schemes to present a baseline against which additional carbon benefits are gained in a predictable and accountable manner through a project is rather difficult (Virgilio and Marshall, 2009); in fact impossible, as in Ghana the data on forest cover and change is so disputed, and poorly founded (Fairhead and Leach, 1998).

History of land-use interventions

The Transition Zone has witnessed a series of forestry interventions within the last couple of decades aimed at managing this dynamic landscape and creating wealth. Prominent among these are private forest plantations, timber concessions and the *Taungya* system, implemented in accordance with the Timber Resources Management Act. The forest plantations entail farmers planting trees, mainly teak, on their own lands and selling off mature trees to timber contractors, whereas timber concessions involve granting Timber Utilisation Contracts (TUCs), specifically to timber contractors who have demonstrated the commitment to provide specific social amenities for affected local communities. Under the *Taungya* system, farmers are given degraded portions of forests to interplant tree seedlings with food crops for the first three years, after which they abandon the land for the trees to mature.

Despite the initial fervour, these forestry interventions later became unpopular because landowners thought the benefits were long-term and not equitable. A new category of external interventions, premised on poverty alleviation through increased agricultural productivity, attractive farm gate prices and short-term income, emerged in the early 1990s. Famous among these were the Cassava for Starch and Sunflower Projects, implemented by the Ministry of Food and Agriculture (MOFA). These projects, however, left farmers in debt because there was no ready market as initially touted. The projects eventually collapsed and were later revived but most landowners became sceptical and apathetic.

Thus, the gloomy history of these forestry and agricultural interventions affected local perceptions, and heightened scepticism about external interventions. It was into this context that the CCP arrived, making its implementation difficult both in technical terms (defining baselines and so gaining access to carbon finance) and in terms of acceptance by local communities (given the repeated failures of past interventions, which to local people looked remarkably similar in aims, content and process).

The Carbon Credit Project

The project was initiated in 2008 by Vision 2050 Forestry Ghana Ltd with technical and financial support from Environmental Development Consultants (EDC) UK and Environment and Rural Development (ERD) Ghana Ltd. The aim was to augment the solution to global climate change through tree planting and voluntary carbon trading. The project was developed and led by Dr. Frank Kofi Frimpong, a private forester and CEO of Vision 2050 Forestry Ghana Ltd.

Grassroots actors and tenure relations

The project mainly engaged landowners. Given the poor history of past interventions, landowners were initially sceptical about the project and it had to rely on the influence of famous media personalities to convince some, while others played the waiting game. Those who were convinced committed their lands to carbon forestry (for at least 20 years) and took full responsibility for maintaining *Cedrela* trees on their farms, expecting carbon credits and agri-environmental benefits in return. The project's contractual agreement allowed the farmer to grow and harvest food crops for as long as possible. The trees and carbon stored belonged to Vision 2050 Forestry and the financiers of the project. Landowners were supposed to receive carbon revenue and land royalties (10 per cent of carbon revenue) for allowing the trees and carbon on their lands.

A sub-project, called the Absentee Farmer Programme (AFP), was simultaneously designed to enhance the inclusion of individuals who desired to participate in the project but did not have easy access to land in the target areas. Under the AFP, vast lands were leased by the project developer from individual landowners for a period of 20 years at various locations in the country to serve as land banks for absentee farmers. Absentees adopted tree plots on the rented landbanks and paid for the costs of maintaining these trees with the ultimate interest of generating cash profits from carbon credits. Some were middle-income professionals, popularly referred to as the 'Very Important Persons' (VIPs), who had financial resources large enough to adopt big tree plantations.

Revaluing forests for carbon

The carbon value of each tree in monetary terms was agreed between farmers and Vision 2050 Forestry to be GH¢100 (US$60).[2] This was for a period of 20 years, justified as the average timeframe for REDD+ contracts globally (Gole, 2011). There was, however, inadequate scientific data to justify the standard carbon value of trees agreed upon. In order to estimate a reliable carbon value, data on carbon benefits (additionality) and project costs are essential (Bond *et al.*, 2011). Carbon stocks of the tree plantations ranged between 231 and 274 tCO_2e per hectare and were expected to increase to approximately 320 tCO_2e per hectare as trees continued to grow (TREES, 2010). However, the lack of baseline data made it impossible to estimate additional carbon benefits from reducing deforestation and degradation. Equally, the total cost of the project could not also be estimated due to lack of quantitative data on opportunity, implementation and transaction costs.

The GH¢100 (US$60) carbon value per tree agreed upon was, therefore, improvised based on the market value of a 20-year-old *Cedrela* tree 'which gives a minimum yield of 5m³ for a negotiable average farm gate price of GH¢75/m³'.[3] This gives a cumulative revenue of GH¢375 (US$225) per tree for 20 years. The remaining GH¢275 (US$165) per tree (after deducting carbon revenue of GH¢100) was crudely projected to cover related implementation and transaction costs. Commenting on the accuracy of the above method of estimating carbon revenue, the project developer said:

> This method, although not free of flaws, is the best we can have, particularly in the absence of sufficient data on historic emission levels and project-related costs. The risk involved is that this method is based on the assumption that potential carbon revenue for *Cedrela* will be more lucrative than its timber revenue, which may not necessarily be the case due to fluctuations in carbon value. The carbon revenue agreed upon may, however, be reviewed with the consent of farmers if the project finally gets access to the voluntary carbon market.[4]

Financial and political challenges

Efforts to solicit funds from voluntary carbon markets and the World Bank's BioCarbon Fund failed as the project was unable to meet the complex and stringent technical requirements for funding. Instead, strategies were adopted to pre-finance the project. One major source of pre-finance was an alleged investment fund of US$750,000 from the EDC UK and ERD Ghana Ltd., a subsidiary of EDC in Ghana.[5] The project was also pre-financed by selectively harvesting and selling teak belonging to the project developer, regardless of the resulting carbon leakage. Membership registration fees paid by farmers were also used to supplement the project's budget. These pre-financing options were, however, not sufficient to develop land banks for absentees and pre-finance landowners as promised.

The financial challenges were exacerbated by alleged political interferences, as narrated by the project developer:

> There is this perception from the ruling government that my project was initiated to advance the political interests of the main opposition party, the NPP. Hence, all the requests for government's support for this project have fallen on deaf ears […] Even there were occasions when I was arrested by the Bureau of National Investigations (the BNI) under the ploy that I had not paid carbon funds to farmers as promised. So I have been wasting many productive hours in the law courts these days at the expense of the project […] Political hindrances are even encountered when thinned out logs have to be exported to pre-finance the project.[6]

Due to the financing challenges, and like the previous land-use interventions, the project collapsed, leaving farmers disappointed, while some VIP absentees sought legal action against the project developer. ERD Ghana Ltd planned to revive the project in 2012 by shifting attention from smallholdings to a plantation model that they judged to be easier to manage and finance, but this has yet to materialize.

Despite its demise, the subsequent sections will show that the CCP was very similar to many other projects being established across the world under the label of carbon forestry for carbon sequestration and REDD. There are some important lessons to be learnt from how the project intervention was framed, by examining the underlying narratives, from various actors.

Narratives: Framing the project intervention

The dominant narratives associated with the project are deeply rooted in assumptions about forest degradation, and its detrimental effects on the climate, agricultural productivity and livelihoods. The need to reforest for carbon sequestration, climate change mitigation, sustainable agriculture and poverty alleviation has been repeatedly underscored. Some of the dominant narratives are in harmony with narratives from the grassroots. Others are, however, challenged by alternative narratives from farmers, including from oral histories. These alternative perspectives underscore the need for opening up policy spaces in the carbon offset policy process to discuss perceptions, challenges and alternatives.

The forest degradation narrative

As discussed earlier, the vegetation of the Transition Zone is a mosaic of forest fragments in the midst of savanna shrubs and grasses. It was commonly narrated by

Vision 2050 Forestry that these forest fragments were remnants of pristine semi-deciduous forests of the past which were being degraded into grassland, hence the need for reforestation. The dominant narrative follows the Ghana Poverty Reduction Strategy's (GPRS) premise that the country's forest cover has reduced from 8.2 million hectares at the beginning of the twentieth century to 1.6 million hectares (Government of Ghana, 2003), a figure repeated frequently in nearly every official document and consultancy report on the subject. Based on an estimated deforestation rate of 65,000 hectares per year, the dominant narrative follows experts' projection that Ghana can lose its entire forest cover in less than 25 years, emphasizing an urgent need for reforestation.

However, the above narratives and deforestation statistics are questionable for a number of reasons. First, the notion that past forests were once in a 'pristine' condition is challenged by oral histories and travellers' accounts. These suggest that the landscape has been a blend of forests and grassland as far back as the nineteenth century (cf. Fairhead and Leach, 1996, for Guinea). The Transition Zone has for centuries been a dynamic system where there has been a cyclical interaction of forest, grassland and fallow in a particular place, generating an overall mosaic. Given such dynamic flux, standard measures of carbon assessment become difficult to use, and simple estimates of forest decline are meaningless.

Forest degradation is popularly blamed on wildfire, particularly the historic 1983 fire. This popular narrative from Vision 2050 Forestry and farmers is reinforced by scientific literature (Hawthorne, 1994; Nsiah-Gyabaah, 1996). An alternative narrative from elderly farmers, however, reveals that it was rather the effects of the fire on cocoa production that resulted in a significant loss of trees from the farming landscape and not the fire in and of itself. Traditional cocoa production which required high-level shade from indigenous trees was the main system of farming prior to the fire. The collapse of cocoa production due to the fire gave way to maize as the dominant crop. Maize production encouraged farmers to remove trees from the farming landscape as it requires an open field for optimum production. An elderly farmer narrated: 'so we needed to burn the trees in our farms to ash in order to grow more maize'.[7]

The dominant narrative also links forest degradation directly to rapid population growth, which increases the demand for farming land (TREES, 2010). In the 'Ghana 2000 and Beyond' document, the World Bank concurs that the land frontier has been receding rapidly in the last three decades, and rising population density has led to the shortening of the fallow period and continuous cropping at the expense of the forest (World Bank, 1993). The World Bank's Country Environment Analysis (CEA) further argues that more than 50 per cent of Ghana's original forest has been converted to agricultural land by slash-and-burn practices (World Bank, 2006).

However, farmer interviews reveal that the extent to which the bush fallow system is practised in the Transition Zone is largely dictated by land ownership and should not be directly attributed to population expansion. Whereas the practice is uncommon or relatively shorter among families with smaller lands, those with customary rights to bigger lands tend to fallow for as long as ten to forty years to rejuvenate the land through shifting cultivation. Some families even practise bush fallow as a means of preserving and rehabilitating degraded farmlands for the future generation. Moreover, it is these old bush fallows that are typically converted to shifting agriculture and not the primary forest as the dominant narrative holds.

In sum, just as it cannot be denied that forest loss has occurred in the Transition Zone over the years, simultaneous gains in vegetation cover through bush fallows and afforestation cannot also be overlooked. Local findings on the long-term existence of grassland vegetation in the Transition Zone; maize production as the main agent of landscape change instead of wildfire; and the establishment of bush fallows and their conversion to shifting agriculture rather than the primary forest, point to several flaws in the dominant framings on forest history. It is therefore imperative to fine-tune the conventional framings with indigenous narratives in order to reconstruct knowledge of past forests. Until this is achieved, the foundation of many carbon projects, which are generally rooted in these dominant framings, will remain disputable.

The food security narrative

Climate change is blamed for dwindling crop yields (Metz *et al.*, 2007); hence, the popular narrative from Vision 2050 Forestry that carbon sequestration through reforestation will ultimately improve agroecological conditions for sustained food production. This storyline is, to some extent, supported by farmers. Farmers narrate that in the first two years of planting, the *Cedrela* trees are able to improve soil fertility, reduce soil compaction and provide shade essential for food crops. The trees have also helped to reduce the incidence of wildfire and its devastating effects on farm produce. As one farmer explained: 'the incidence of wildfire on our farms has reduced significantly because we have been putting in extra effort to protect the trees from fire'.[8]

The dominant narrative on food security is, however, challenged, as the farmers point out that the *Cedrela* canopy inhibits crop production after two years of planting. Some farmers did not fully grasp the project concept of switching from agroforestry to tree plantation at the consultation phase of the project, hence their concerns on losing food crops. The carbon project therefore requires farmers to integrate trees in new ways in the farms. The project design only works for those with large land areas, where forested patches and food production areas can be separated. For others, the basic imperative of food production is paramount, and the carbon project is incompatible after two years of *Cedrela* growth.

The climate change narrative

From the perspective of Vision 2050 Forestry, carbon emissions from forest degradation are a major cause of climate change, hence the justification to reclaim 'degraded forests' by planting trees for carbon sequestration, carbon credits and ultimately climate change mitigation. This narrative, allied to the international scientific and policy consensus, was generally shared by VIP absentee farmers who had some knowledge from formal sources of environmental issues and good access to media reports on climate change.

Farmers, however, view trees more in relation to micro-scale weather systems offering stability in rainfall patterns, and protection against erratic seasons and wildfire risk. Most of them had never heard of carbon as a constituent element of trees, let alone the idea that it could be sold. They draw their understanding of carbon as a commodity primarily from the explanations given by Vision 2050 Forestry during sensitization workshops. Through these, farmers have come to perceive carbon as, 'money locked up in the air that can only be accessed by planting trees to tap the money'.[9] They have also come to understand that, 'for every aircraft that flies an amount of money, linked to exhaust gases, is saved up in the air for the tree planter'.[10]

Even so, the link to carbon remains alien, and not clearly understood. They are sceptical about the carbon sequestration project, arguing that the benefits lie in tree planting, a practice they know well. Also, for them, the project was no different from past interventions focused on community and individual farm-based forestry, and their memories of past failures are widely shared. As generations before, farmers see benefits in tree planting in certain places on the farm, for specific reasons and as part of a wider farming and livelihood strategy, with trees intimately linked in. Trees are part of lived-in landscapes with long histories, with certain trees marking grave sites and past village settlements, and some being preserved as 'sacred groves' protected by the spirits of the ancestors.

The idea of carbon sequestration is therefore not part of this worldview, and the range of policy narratives that justify the project intervention had little purchase at the local level. There was thus a mismatch between perspectives; a disjuncture that undermined the project's ability to work effectively on the ground.

The job creation, poverty alleviation and youth migration narrative

Planting trees for carbon credits was touted as a sustainable income-generating venture and pro-poor policy intervention, especially in the rural areas where the main source of livelihood is farming (Vision 2050 Forestry, n.d.). Based on a booming carbon market scenario, the dominant narrative further held that carbon revenue could serve as enough incentive to curb rural–urban migration of the youth. The slogan 'plant trees for cash' was accordingly used to drive home the perceived financial prospects of the project to the general public.

The different categories of farmers held different views on the financial prospects of the project. Absentees generally considered the carbon revenue as economically viable since the project was a part-time business that took little toll on their main livelihoods. Landowners, however, considered the carbon revenue to be inadequate in view of inflation and the high opportunity cost of farm land, but were enticed by a pre-finance package that was expected to be available shortly after signing up for the project. Landowners did not feel the alleviation of poverty because the pre-finance package was not paid as promised. For the project to be regarded as pro-poor, landowners are of the opinion that carbon revenue must outweigh income from highly profitable crops such as chili pepper, cocoa, maize and marijuana, and has to be paid on a short-term basis. Landowners found these alternative crops to be more lucrative, thus reducing incentives to engage in the project.

Moreover, youth involvement in the project was found to be lacking. The reasons for youth apathy are complex; they go beyond urbanization and vary between the peri-urban and rural areas. In Dumasua and its environs, which are peri-urban, the youth are attracted to lucrative job opportunities in Sunyani, the nearby regional capital, and generally tend to associate farming with failure. Youth involvement was relatively encouraging in Badu and its suburbs, which are rural, due to limited availability of alternative jobs to farming. The long-term investment return of the project was, however, a disincentive to a section of the youth who were interested in short-term profits. This category of youth therefore preferred to engage illegally in marijuana farming for relatively high and short-term profits regardless of the risks involved.[11]

Thus, in contrast to the simple narrative that carbon forestry will result in pro-poor gains, especially for the youth, the situation on the ground is much more complex. Farmers highlight the relative returns to carbon forestry versus other opportunities. These vary across sites, with urban employment being attractive in Dumasua, while marijuana farming is a good earner in Badu. The time frame of returns is also significant, with farmers emphasizing the importance of immediate, short-term returns in contrast to the long-term prospects promised under carbon forestry. Carbon forestry, therefore, enters a highly differentiated setting where its impacts are conditioned by a range of factors not considered as part of the 'pro-poor' wealth and job creation narrative promulgated by Vision 2050 and others.

Impacts on resource access

In further critiquing the 'pro-poor' and job creation storyline it is important to look at how the carbon project interacted with resource access and agricultural livelihoods, including how farming systems are being shaped.

Although the CPP had many challenges, the project was certainly not without certain benefits. It is therefore necessary to examine how different processes affected local actors' ability to join the project and derive benefits from it, focusing on the power relations affecting inclusion and exclusion. In the following sections, particular attention is paid to socially differentiated patterns of access, drawing out comparisons between different actor groups in terms of land ownership, ethnicity, gender, wealth, age and the nature of livelihood.

Land ownership: Landowners versus migrants

In the CCP, land ownership primarily dictated the extent of inclusion of small-holders and what benefits could be derived. Access to farming land is primarily guaranteed through kinship and membership of a particular clan. The existing land tenure system means that migrants cannot own land, but have to depend on other arrangements such as sharecropping and land renting to acquire land. The lack of legitimate rights to land inheritance therefore excluded many migrants from the project.

It was against this background that the AFP was initiated to encourage the participation of the landless. This notwithstanding, legitimate landowners enjoyed certain property rights over absentees, such as the right to food crops, NTFPs and land royalties. Land ownership also offers better security against the collapse of the project. One landowner narrated, 'we have the right to sell the trees when they mature or leave them as legacy for our children now that the project has collapsed because the trees are standing on our own lands'.[12]

Ethnicity: Migrants versus indigenes

Although the Absentee Farmer Programme (AFP) of the carbon project aimed to enhance the inclusion of the landless, land access shaped by ethnicity still presented barriers to the inclusion of non-indigenes. Migrants likened the AFP to weak land rental agreements that have existed in the communities long before the carbon project was introduced. Landowners are often compelled by emergency situations to rent out their lands and, after settling emergency bills, some covetous landlords, motivated by the absence of enforceable sanctions, decide to abrogate the land rental agreement by paying off the tenant farmer. This often leads to a land dispute as the tenant farmer will strive not to lose investments made on the land to the landowner.

Non-indigenes therefore reason that land rental agreements under the AFP cannot similarly offer them sufficient security to the land to justify long-term investment in carbon offset. Hence, local migrants were rarely involved in the AFP and for that matter the CCP. The majority of absentees were non-residents who were less concerned with local land rental issues.

Gender: Men versus women

Traditional inheritance rules and norms often exclude women from inheriting and controlling land. The lands are customarily for the males in the family because they supposedly have more physical strength to make optimum use of the land. This gender inequity in land control and distribution generally gives men better access to land and bigger holding sizes than women (FAO, 2011). For this reason, few women were involved in the CCP. Moreover, big landholding sizes allowed men generally to plant more trees than women and also enabled them to cope better with the competition between trees and crops by committing surplus land to the project.

Wealth: Rich versus the poor

The financial challenges of the CCP necessitated monetary commitments from farmers in the form of registration and maintenance fees that were not affordable to the poor, thus leading to their exclusion from the project. One farmer observed, 'I know of people who could not join the project simply because they could not afford the registration fees required'.[13] Some farmers even had to borrow money or sell their property in order to afford project-related costs. Thus, the project generally favoured the inclusion of wealthier actors, including wealthy VIP absentees who were able to adopt large tree plots, thereby undermining the 'pro-poor' claims.

Age: The elderly versus young people

Traditionally, fathers control lands on behalf of the entire family. Mature male children get portions of the land usually when fathers have grown very old and are not strong enough to manage the entire area. Thus, in a situation where there are many mature male children, each gets just a small portion of the land. Old people's better access to land enabled them to voluntarily plant additional trees, beyond what was required by the carbon project, for extra economic gains, in contrast with the youth.

Nature of livelihood

The advent of CCP has indirectly been excluding hunters and charcoal burners, whose operations often result in uncontrolled burning of farms. Farmers used to force out fire users from their farms even before the CCP arrived, but to a limited extent. The carbon project has only made farmers more cautious about anthropogenic causes of wildfire, since the trees have added value to farmlands. Many farmers, therefore, have to guard their farms more frequently against fire users as a measure to protect their trees, including drawing on the services of the military and the Fire Volunteer Squad.

Overall, the claims that the project was pro-poor, supporting local communities through investment, can be questioned. A highly differentiated impact is observed, with many groups excluded. Exclusion occurs by force when hunters and charcoal burners are chased from crop lands.

In sum, older, male land owners with large farms are the dominant group involved in the project. Rights over land are legitimated by customary inheritance systems, allowing landowners to exclude others. Rights to land are thus an important factor influencing inclusion and exclusion. Until customary land laws become free of kinship, gender and age discrepancies, the inclusion of migrants, women and young people in forest carbon interventions will continue to be restricted.

Impacts on livelihoods

Being a land-use intervention, the carbon project certainly had significant implications on various livelihoods. This section examines how the livelihoods of subsistence farmers, NTFP collectors, sharecroppers and migrant farm labourers are influenced by the CCP.

Small-scale farmers/NTFP collectors

As discussed earlier, farmers have found the *Cedrela* trees prevent fires, improve crop yield and hold the potential for NTFP commercialization. Some farmers are, however, concerned about losing food crops, their main source of livelihood, when the tree canopies close. These concerns with food security, however, only came from Dumasua (a peri-urban site) where land is relatively scarce and expensive due to urban expansion and a growing demand for building, and specifically from farmers with limited lands who were mostly women. Older male farmers with better access to land have been able to develop effective coping mechanisms. Those who owned multiple farms usually committed only one farm to carbon offset and reserved the rest solely for crop production. Those who have single but large farms would retire only a small portion for carbon offset, leaving the rest for food crops.

Migrant farm labourers

Hiring the services of migrant farm labourers is one significant way the CCP is contributing to job creation. Farm labourers are regularly hired to weed around tree seedlings and prune mature trees. The livelihoods of these farm labourers depend largely on hiring by older farmers and as a response to the expansion of invasive weeds. A farm labourer on average earns GH¢15 from weeding for only three hours in a day (usually from 9am till 12 noon), which is about three times the daily minimum wage of GH¢5.24 in Ghana.

Sharecroppers

Sharecropping arrangements are alternative ways the landless, especially migrant farmers, can have access to farmlands in order to earn a living. Ideally, a sharecropping arrangement is a win–win situation for both the landowner and the sharecropper. However, because it is difficult for landowners to monitor production from the land, there is often distrust between landowners and sharecroppers. Some dissatisfied landowners are motivated by the 'plant trees for cash' idea and the fact that trees are easier to monitor than food crops, and are therefore considering investing their lands in trees rather than in sharecropping. An old woman complained:

> My sharecropper is a cheat; he takes almost all the farm produce and gives me just a little because he knows am I not strong enough to monitor his activities. If only you [referring to the researcher] can offer me additional tree seedlings, I will sack him and use the whole land for trees since it will be easy to monitor the trees.[14]

Thus, sharecroppers stand a risk of being displaced by trees, especially considering the fact that sharecropping arrangements do not exist for tree plantations. Sharecroppers will not be displaced alone, but together with food crops, and with adverse implications on local food security.

In sum, the carbon project has both positive and negative livelihood implications. The *Cedrela* trees present an opportunity for farmers with large land areas to diversify income through agroforestry and NTFPs, but those of limited lands have difficulty coping with the competition between trees and crops. While the project is promoting the livelihoods of migrant farm labourers, the potential displacement of crops by trees will have negative consequences on the livelihoods of sharecroppers.

Impacts on farming landscapes

Besides increasing the tree stock of farmlands, the introduction of the carbon offset trees is also triggering a shift from food crop to cocoa production. As discussed earlier, cocoa was once cultivated on a large scale in the Transition Zone until the 1983 bush fires. Farmers had the desire to grow cocoa again long before the carbon project, because of the attractive market for cocoa and the introduction of early-bearing and high-yielding hybrid varieties. The fear of wildfire had, nonetheless, stood between farmers and cocoa farming. The introduction of the *Cedrela* trees has given farmers the hope that wildfire can be prevented to some extent, and this is encouraging many to start growing cocoa. Some cocoa has already been established close to tree plantations where the likelihood of wildfire is thought to be low.

Farmers are, however, now not cultivating the traditional shade-requiring cocoa variety, but high-yielding hybrid cocoa which can grow in full sun, and requires no shade trees. This is encouraging farmers to eliminate shade trees from the farming landscape in order to intensify production. Thus, the 'carbon trees' may likewise have to give way to cocoa, just as earlier maize became an agent of deforestation. This is having a net negative impact on carbon sequestration at the wider landscape level.

Hybrid cocoa appeals more to landowners and the youth since financial returns are higher and short-term as compared to carbon revenue, thus undermining the prospects of investing in carbon in farm lands. Intensification of hybrid cocoa, just like shifting from food crops to tree plantations, is also limiting the availability of farmland for food crop production. This can have adverse consequences on food security, but farmers are optimistic of achieving livelihood security through cocoa income.

Conclusion

The experience of the CCP in Ghana is an important reminder of the challenges of realizing the ambitious goals projected in global narratives about the new carbon economy. While the project was not compatible with the REDD framework in Ghana, and would have been rejected on multiple fronts for technical non-compliance, the CCP is not dissimilar to many projects being elaborated across rural Africa in the hope of tapping into what are assumed to be highly lucrative carbon finance options (see other chapters in this book).

The CCP story will likely be the fate of many other REDD+-type projects, as high expectations fail to match the rigorous requirements for gaining carbon finance. In the end, many have no prospect of delivering and, as with the CCP, the business model of harvesting trees to pre-finance the project, despite the resulting carbon leakages, became inevitable, given the strict technical requirements to raise formal carbon finance. The CCP story suggests that, in many cases, the prospects of realizing the 'triple win' claims of the global narrative will remain a mirage, and instead a 'carbon grab', where land is leased in large blocks managed as a plantation, is the more likely outcome.

Studying project failure is perhaps a bit depressing – and many aspects of the CCP were certainly disappointing and have only exacerbated the negative experience of past land-use interventions. But it is essential to learn lessons from these experiences. The CPP case suggests the following conclusions.

First, the justifications for gaining access to carbon markets depend on establishing clear baselines and data that shows additionality. This has proved exceptionally difficult because of the technical challenges of measurement and problems with existing baseline data on forest cover and deforestation rates. National assessments of forest change are notoriously unreliable, undermining claims made for forest carbon projects. Local understandings of forest change emphasize dynamic change rather than secular decline, and past and current forest interventions in the transition zone are about managing this dynamic. The requirements for entry into formal carbon markets and REDD-type project funding are so stringent it is unlikely that smallholders in the project area will ever be eligible. An alternative route of government finance, following receipt of international payments based on a general reduction in carbon emissions and allocating payments to carbon projects at the local level, is probably the only route to engage smallholders in such efforts, if the CCP experience is anything to go by (Van Bodegom et al., 2009). The alternative, as proposed now by Vision 2050, is a plantation-style forestry option, which has a poor track record in Ghana, and limited benefits for local farmers.

Second, the justification for carbon offsets in smallholder contexts should be based more on farmers' understanding of climate change and livelihood concerns. Farmers lack understanding of carbon offset mechanisms and can rarely make connections between carbon and climate change. Instead, their understanding of climate change is primarily based on the effects of apparent environmental changes and local weather systems on their livelihoods. Tree planting to improve rainfall, fire prevention and availability of NTFPs is welcomed, and encourages participation of farmers and landowners, while arguments about carbon sequestration remain alien.

Third, with a long-term financial return of an uncertain amount, dependent on project finance and access to carbon markets, carbon offset options must compete with other land uses. In this case, high value crops offered a higher return over a shorter period, reducing incentives to engage with the project. Land value is increasing, especially in peri-urban sites where demand for land for building is growing. These areas are unlikely to offer good opportunities for carbon investments, as returns from other options will be higher. It is only in relatively land-rich areas, where farms are in large blocks and new land uses are not competing with other uses, that carbon forestry is likely to take off. But questions are raised about the future, given pressures on land through growing populations and expanding urban areas. Carbon offset schemes are designed to operate over a minimum of 20 years, and it is clear that land values will change dramatically over this period, potentially undermining the incentives to maintain carbon trees, even in areas that have land surplus now.

Fourth, as in every other form of land use, it is not exceptional for carbon offset interventions to bring about some form of exclusion. Hunters and charcoal burners have been excluded from areas under carbon forestry, undermining their livelihood options. Land inheritance systems dictate the degree of inclusion or exclusion and the extent to which benefits from carbon offsets can be accessed. Until customary tenure arrangements become free of kinship, gender and age discrepancies, the inclusion and resource access of migrants, women and young people in carbon offset interventions will continue to be restricted.

Fifth, changes in farming landscapes brought about by carbon offset interventions can have both positive and negative livelihood implications. The *Cedrela* trees present an opportunity for farmers with large land areas to diversify income through agroforestry and NTFP commercialization, but those with limited land have difficulty coping with the competition between trees and crops. Although achieving the purposes of food crop production and carbon storage on the same piece of land is possible, it is necessary to increase farmers' awareness of possible trade-offs and to customize the project design to suit the land availability in the area and of individual farmers. While the introduction of trees into the farming landscape is promoting the livelihoods of migrant farm labourers through the generation of employment, the potential displacement of crops by trees will have negative consequences on the livelihoods of sharecroppers. The potential shift from food crops to trees and the intensification of hybrid cocoa will also have adverse implications on food security and carbon forestry.

In sum, the CCP case study has therefore highlighted some of the dilemmas presented by the enthusiastic rush to carbon offset projects as part of agricultural and rural development, under REDD+-type programmes. If not implemented effectively they can result in insecure carbon sequestration (due to lack of permanence and leakage), limited finance flow (due to compliance challenges), the exclusion of some people within communities (in this case mostly poorer farmers, young people and women, as well as migrants and sharecroppers) and benefits accumulating to a limited group (in this case mostly male farmers with larger land areas and some absentee investors). A carbon forestry project can result in declines in food production and displacement of livelihoods (including, in this case, hunters and charcoal burners). Despite the rhetoric, carbon offset projects outcomes are therefore not automatically 'pro-poor' and 'win–win'. This depends on power relations within local communities that can include and exclude, as affected by gender, ethnicity, migrancy, age, wealth and land ownership rights. Unless such issues are taken into account in project design and implementation, carbon projects will inevitably have unequal and uncertain outcomes.

Notes

1 For a longer and more detailed overview of the project and its impacts, see Hashmiu (2012).
2 All exchange rates are based on November 2011 rate of US$1≈GH¢1.67.
3 Interview: Dr. Frank Kofi Frimpong, Tainso, November 2011.
4 Interview: Dr. Frank Kofi Frimpong, Tainso, November 2011.
5 Dr. Ed Yeboah Bossman (CEO of ERD Ghana Ltd): Personal Communication.
6 Interview: Dr. Frank Kofi Frimpong, Tainso, October 2011.
7 Transect-walk conversation: farmer, Badu, March 2011.
8 Focus group discussion: farmer, Dumasua, February 2012.
9 Focus group discussions: farmer, Badu, October 2011.
10 Focus group discussions: farmer, Badu, October 2011.
11 See Hashmiu (2012) for a comparative economic analysis of the different land-use options.
12 Transect walk conversation: landowner, Dumasua, March 2014.
13 Focus group discussion: farmer, Dumasua, September 2011.
14 Focus group discussion: landowner, Badu, March 2012.

10

OLD RESERVE, NEW CARBON INTERESTS

The case of the Western Area Peninsula forest
(WAPFoR), Sierra Leone

Thomas Winnebah and Melissa Leach

Introduction

The last few years have witnessed growing enthusiasm for carbon forestry in Sierra
Leone. Developments are still at a relatively early stage, with just a handful of
projects being planned or implemented. Yet in the context of international policy
interest and perceived market opportunities, government, Non Governmental
Organizations (NGOs) and private sector actors alike now see selling carbon credits
as a promising new way to generate finance for forest conservation and to link
forestry with profit. One of the first schemes, and the focus of this chapter, is the
Western Area Peninsula Forest (WAPFoR) carbon project.

The project is situated in the dense humid forests that cover about 17,600
hectares of the hills bordering the expanding coastal capital city of Freetown. Sierra
Leone's first forest reserve was established here in 1916 by the British colonial
administration primarily to 'protect them from farmers in order to preserve their
industrial purpose' (SAH/D, 2007, p.31) and from particularly 'Temne sawyers
and canoe makers' (Munro, 2009, p.106). The reserve, formally defined as
government-owned land, has subsequently experienced a complex history and a
succession of conservation initiatives with different aims, implemented by different
institutions. Carbon is the latest dimension in this landscape and intervention
history. In 2009, the five year Conservation of the Sierra Leonean WAPFoR and
its Watershed project implemented by the Forestry Division (FD) of the Ministry
of Agriculture, Forestry and Food Security (MAFFS), Welthungerhilfe (WHH)
and the NGO network Environmental Forum for Action (ENFORAC) was
established with 80 per cent European Union (EU) funding and 20 per cent WHH
funding. The aim was to reinforce protection of the reserve and promote its
sustainable use. For continuity after the end of EU funding in February 2014,
WHH had been developing a Reduced Emissions from Deforestation and

Degradation plus (REDD+) programme geared to voluntary carbon markets as an alternative source of financing. In 2013, MAFFS and WHH sought funding and partnerships to finalize a Project Description and progress towards implementation of an intended 'pro-poor' carbon forestry scheme. MAFFS 'started implementing a National REDD+ and capacity building project with funding from the EU in March 2014' (personal interview, MAFFS REDD+, June 2014; RoSL, MAFFS and EU, 2014), but 'how far will the government get WAPFoR into the EU REDD+ project is a major threat' (personal interview, WHH, Freetown, 24 October 2013).

The chapter explores the intersection of this long-established forest reserve landscape, with new carbon interests. It draws on a series of key informant interviews and focus group discussions with villagers of different ages and genders, and resident and immigrant status, community mapping, proportional piling and transect walks in six of the WAPFoR's directly targeted[1] 30 villages surrounding the reserve, as well as on personal interviews with government, NGO and project staff and forest guards in 2013. The chapter addresses the WAPFoR's complex history, the ways this has been drawn on in establishing the carbon project, and how the project is perceived and experienced by villagers in the context of their varied livelihood interests and activities. In particular, it reveals two paradoxes. First, despite its embeddedness in a long landscape history, the carbon project has had to rework that history into a series of narratives in order to justify itself. Second, despite the project's aims and claims to be 'pro-poor' and 'participatory', the complex of institutional arrangements, tenure relations, ideas and imaginaries that it relies on works against realizing these aims in practice. It has proved difficult to reconcile the development of the carbon project with local perspectives and priorities. Instead, for most local community members, the project is experienced as a continuation of the fortress forest conservation of the past, now reinforced more strongly than ever and hedged around with new, hard-to-fathom ideas around carbon and money.

The Western Area Peninsula forest: Landscape and history

Located about 5km south of Freetown in the Western Area/Region of Sierra Leone, the WAPFoR covers a narrow elongated chain of hills approximately 37km long and 14km wide, with a range of peaks, the highest being Picket Hill in the south, which rises to about 900m (Forestry Division, 2013). Dense humid forest is supported by high rainfall (3,000–7,000mm per annum). The rainfall supports a range of streams, rivers and springs that supply water to surrounding areas.

To the north, the WAPFoR borders the expanding capital city of Freetown, while to the south, west and east lie narrow coastal plains inhabited by rural and peri-urban communities, with livelihoods based mainly on fishing and farming respectively. Since pre-colonial times the area has been inhabited by Temne and Sherbro people, who were joined by Krio settlers after the abolition of slavery from the late 1700s. The sites of pre-colonial Temne settlements can still be identified from their vegetational markers within the forest. As an elderly man explained, 'our ancestors owned the forest without knowing that there was a boundary. They grew ginger and cassava, which they sold in Freetown. There was no one in-charge of there' (interview, Kossoh Town, 29 March 2013).

In 1916 and following a series of interventions from British colonial foresters, A. H. Unwin and C. E. Lane-Poole, the bulk of the closed forest area was demarcated as the WAPFoR reserve (henceforth referred to as the reserve) and claimed as government-owned land, within the British Crown Colony established on the Western Area peninsula (Munro, 2009). As the former Headman of Bonga Wharf put it, 'the white man went ahead and demarcated the boundary on their own' (focus group discussion, Bonga Wharf, March 2013). Forest reservation at this time purportedly focused on protecting a key watershed vital to the water supplies of the colony and its growing capital city, in the context of the expanding European timber industry for shipbuilding and other purposes (SAH/D, 2007; Munro, 2009). Narratives about the degrading effects of so-called shifting cultivation – the bush-fallow system practiced by Temne and Sherbro farmers – also took hold from this time, drawing on widespread colonial scientific discourses. Munro (2009, p.106) gave account of 'the colonial government employing two "experts" from England', who in their subsequent reports 'noted areas of forest destruction on the Peninsula, placing blame directly on the inefficient and destructive activities of the native population; with a particular emphasis on the Temne sawyers and canoe makers'.

From the earliest, the reserve's proximity to Freetown and visual prominence from the city and the coast gave it a public aesthetic and political importance. Indeed, the colonial core forest reserve boundary had 'an extension [...] demarcated by the Forestry Division in 1961 partly for protection of the Freetown water supply catchment and to improve the scenic view of the landscape' (personal interview, MAFFS, March 2013).

Government ownership and management of the forest reserve continued after Independence. Watershed protection continued as a priority, and several rivers within the reserve were dammed to secure water supplies to Freetown's growing population. However, in the context of growing international interest, biodiversity became an additional focus for conservation. To protect the reserve's distinct fauna (more than 374 species of birds, and charismatic species such as the white-necked Picathartes bird and a variety of apes), it was declared a statutory non-hunting reserve in 1972. In addition, the statute gave the Head of the FD authority to grant license concessions to people and investment entities to use parts of it for farming, stone quarrying and exploitation of mineral resources. The jurisdiction enhanced its conversion into private property and encroachment, as put by a government official: 'Forestry started the whole thing by giving license to do farming annually [...] They will in turn sell the lands' (personal interview, Freetown, 27 September 2013). Private and NGO biodiversity conservation activities were initiated, including efforts by the Sierra Leone Conservation Society (SLCS) and the establishment of the Tacugama Chimpanzee Sanctuary (TCS) within the forest. Increasingly, the reserve's distinct and high biodiversity (Karim *et al.*, 2013) was highlighted in justifying it as one of the country's and sub-region's protected areas – the westernmost extent of the Western Guinean lowland rainforest in West Africa (Forestry Division, 2013).

The country's socio-economic and governance system deteriorated during the decade after 1980. The donor-driven Structural Adjustment Programme (SAP) led to increasing waves of rural–urban migrants (e.g. education and job seekers, 'hypermobile impoverished youths', deprived local people and later Liberian refugees) into Freetown (Winnebah, 2003; SAH/D, 2007; Higuchi, 2009; Davies, 2012; Fanthorpe and Gabelle, 2013). As Freetown expanded, the reserve's land resources came under increasing pressure for constructing dwelling houses and institutional offices, and timber, charcoal and stones to meet the building and everyday needs of growing the urban populations. These pressures intensified during and in the aftermath of Sierra Leone's 1991–2002 civil war, when many Internally Displaced Persons (IDPs) and refugees from the hinterland and the Republic of Liberia respectively, flocked to the peri-urban and rural areas in the eastern and southern parts of the reserve near Freetown (Winnebah and Cofie, 2007). In addition, huge waves of migrants from Freetown sought safe haven in the east, north and west of the reserve, after the Revolutionary United Front (RUF) launched a direct attack on Freetown from there in January 1999 (SAH/D, 2007; Munro, 2009).

This led to significant illegal forest clearance around old peri-urban villages, such as Grafton-Jui-Kossoh Town in the east, Hill Station-Regent in the north and Lumley-Goderich in the west, for settlement near and within the reserve (EuropeAid/126201/C/ACT/Multi, 2008). But most houses built there are owned by relatively wealthy Freetown residents looking for better security, space and views from the city's periphery. Major forest clearance and the opening of space for such residential development was also hastened by the construction of the British International Military Advisory Assistance and Training (IMAAT) Centre, the subsequent construction of the office of the EU and the relocation of the American Embassy from Freetown to the Hill Station/Leicester Peak area for security reasons after the end of the war in 2002. This triggered massive housing expansion in those areas by wealthy elites, some of whom rented or leased their buildings to mostly international agencies and organizations. As a government official put it, 'most of the internally-displaced people are now landlords and in the land business, where they brush and sell land' (personal interview, Freetown, 27 October 2013).

A dynamic set in, locally termed 'land grabbing', whereby typically groups of immigrant youths are sent by Freetown residents to 'brush' and burn parts of the reserve's lands at night, when forest guards are off-shift. The next steps are to profit by making charcoal from the felled vegetation; to put a makeshift shack on the land as a marker, lived in by a poorer client, and to seek a paper declaration from the Ministry of Lands, Country Planning and the Environment (MLCPE) to legitimize the holding, before going on to develop it. Ambiguity and political contestation has since opened up between MAFFS and the MLCPE over their respective rights and responsibilities around such land claims. While MLCPE claim that this is legitimate urban development, MAFFS accuse MLCPE officials of engaging in an illegal process that is damaging forest conservation. Despite the ongoing implementation of the Local Council Act of 2004, local municipal authorities – the Freetown City Council (FCC) and the Western Area Rural District Council (WARDC) – appear to lack a mandate to intervene.

An intense 'blame game' (Munro, 2009) unfolded. While some community people and MAFFS Forestry officials blame MLCPE officials for condoning 'land grabbing', the latter equally blame the former, accusing forest guards of conniving with illegal settlers. Community members are also involved. Thus, government officials accuse villagers and their chiefs of conniving with illegal settlers and loggers in encouraging reserve encroachment. Chiefs and community members for their part justify these actions by drawing on evidence of old pre-colonial settlements inside the reserve to argue that this is ancient community farmland over which they have rights. One consequence of these processes was the contraction of the overall forest reserve by 32 per cent since its original 1916 demarcation (EuropeAid/126201/C/ACT/Multi, 2008).

Thus, by the early 2000s the FD of MAFFS and its meagre force of forest guards struggled to control conversion of and encroachment into the reserve. From 2004, the Director of MAFFS' FD and ENFORAC began to advocate for a major, externally funded forest conservation project. Long-established concerns with watershed protection, and to a lesser extent biodiversity, provided the main justification. Indeed, in 2008, a study of Freetown's water supply underlined the forest's importance and enabled MAFFS to approach the Office of National Security (ONS) on the grounds that forest conservation was a matter of national (water) security. In this context, the WAPFoR project was established and commenced operation in 2009. From the outset, water was central to the WAPFoR project's aims, political value and public image with the project's slogan, '*wata en forest na life*' ('water in the forest is life') adorning numerous brightly painted signs dotted around the reserve. A specific objective was to protect the Guma Valley and the Congo Dams, which supplies water to the entire population of Freetown – by now around 1.5 million people (Moninger, 2014).

The project took an established 'conservation with development' approach. A strict core reserve was re-established to protect water supplies and biodiversity, redrawing and clearly demarcating the boundary around a smaller area to exclude land now converted to urban settlement. To protect the reserve, additional forest guards were recruited – some from nearby communities – trained and partly equipped. These efforts were further strengthened with a successful donor, government and NGO campaign to have the forest reserve upgraded to national park status, becoming the statutory Western Area Peninsula National Park (WAP-NP) in November 2012 (Moninger, 2014). Certain measures have been put in place to protect, maintain and sustain the re-demarcated boundary. For example, the number of forest guards was increased from 16 in 2009 to 36 and 42 in 2011 and 2013 respectively, and boosted by the assistance of 13 armed personnel from the Sierra Leone Police's Operational Support Division (SLP OSD) in 2012 (Moninger, 2014). The Environmental Protection Agency, Sierra Leone (EPA-SL) became involved in subsidizing the erection of boundary posts; monthly remuneration for forest guards and patrolling of the new reserve area; and erecting perimeter fencing in the demarcated areas. Community liaison officers/mobilizers and Environmental Management Committees (EMCs) were established with the aim of raising awareness and interest in maintaining the new reserve boundary, including joint patrolling.

Despite this intensified protection of a forest 'fortress', however, there were ongoing institutional and logistical challenges in sustaining the new boundary. Forest guards received bicycles, rain gear and mobile phones to assist patrols, but struggled to maintain and replace these when they wore out. The SLP OSD complained about lack of assigned open truck transportation for patrolling in large numbers for safety against well-armed offenders. The EMCs complained that they lack logistics to work alongside the forest guards and SLP OSDs. Meanwhile, the protracted conflict of interests between government Ministries, Departments and Agencies (MDAs) over urban land expansion continues to pose serious challenges to the completion, maintenance and sustainability of the new boundary (Moninger, 2014). Thus, as the former headman of Bonga Wharf indicated, even where the re-demarcation of the reserve has been carried out 'there is land grabbing. People are breaking the beacons and encroaching' (interview, Bonga Wharf, March 2013).

Around the reserve a 150m buffer zone was established in which community-led planting and management of fuel wood lots, timber and fruit trees (e.g. guava, oranges, lime, cashew, pear, mango, tombi) was carried out. Alternative livelihood activities and conservation education were implemented in the 30 targeted communities surrounding the reserve. These activities were partly intended to boost community acceptance of the forest conservation activities, in line with a project approach that was claimed to be 'pro-poor' and 'participatory'. However, they were also seen as means to economize and substitute for local resource uses that would otherwise degrade forest within the reserve and its buffer zone.

Indeed, despite the dynamic described above, in which most pressures on the reserve came from urban expansion and some logging and stone quarrying by immigrants and outsiders, the project took a view that attributed significant deforestation to small-scale farming, fuelwood harvesting and charcoal burning by impoverished communities (Moninger, 2011). Evidence for this was limited; surveys undertaken at the start of the project revealed virtually no forest loss around the northern, eastern and southern reserve boundaries. Indeed, there were some areas of forest expansion, likely due to families moving out of farming and into fishing or tourism, and so abandoning bush fallow to grow up into secondary forest. Most communities in recent years have depended on fishing and tourism, and can accommodate a limited amount of farming and vegetable gardening in the areas of land between settlements and the reserve. Nevertheless, this narrative of deforestation by shifting cultivators and 'woodcutters, charcoal burners and stone miners', the actors stigmatized on signboards, became important in justifying the

WAPFoR's 'conservation with development' project. They became even more significant as it turned to carbon.

The WAPFoR carbon project

With the project's EU funding due to finish in February 2014, its implementers were keen to find an alternative, sustainable source of finance to ensure forest conservation for the long term. Encouraged particularly by the German WHH Project Manager, it sought this through carbon finance, initiating the preparation of a REDD+ project in 2011.

The envisaged carbon project (henceforth referred to as the project) was an avoided deforestation project to be registered under Verified Carbon Standard (VCS). A new overall objective for the now carbon-focused WAPFoR project was created (WHH, 2014, p.1): 'to reduce emissions from deforestation and degradation in the Western Area Peninsula Forest Reserve'. Within this, continuity prevailed, with most of the original WAPFoR project objectives unchanged: viz. increasing the protection status from a non-hunting forest reserve to a National Park and a World Heritage Site; setting up an effective, financially autonomous park management and administration; improved patrolling and law enforcement; eradicating urban sprawl and encroachment into the reserve; supporting alternative income-generating activities for communities living along the reserve's boundaries and offering recreational, educational and tourism activities for international and domestic visitors in the area. The project continued with these activities.

It strengthened them further, rectified past inadequacies, and sustained them into the future through carbon finance. Moreover, these activities were reworked as a means to the overall end of reducing carbon emissions from deforestation. In the project's publicity brochure, other conservation objectives which previously held priority status – protecting water supplies and biodiversity – were redefined as additional spin-offs, or 'co-benefits'. Furthermore, in line with the new aim of tapping into emerging carbon markets, two additional objectives were added: setting-up payment for environmental service schemes, especially for climate mitigation and water supply; and promoting offset and sponsoring programmes for the private sector.

By the end of the WAPFoR project in February 2014, the project had passed its pre-feasibility stage and was 'looking for investors interested in supporting further project development according to VCS standards, and/or buying future carbon credits' (WHH, 2014, p.1). The project brochure invited investment partners for what was pitched as a profitable activity. The WHH Project Director, the main developer, envisaged future purchases of carbon credits not just from organizations seeking to offset emissions, but also from mining companies and other private sector actors operating in Sierra Leone's pro-foreign investment climate that may want to green their image and meet corporate social and environmental responsibility targets.

The project developers undertook several carbon project preparatory actions consistent with VCS requirements. In 2011, the Austrian consultancy firm

Österreichische Bundesforste (OBF) was commissioned to conduct a series of scoping studies. Their detailed report and annexes (OBF, 2012) included a Project Idea Note (PIN) that recommended proceeding with the development of a full Project Design Document (PDD), and thence to project implementation. The VCS methodology VM0015 ('avoided unplanned deforestation') (VCS, 2013a) was used for the scoping study and project development documentation prepared by OBF (2012). In applying this, the consultants made a number of assumptions and choices that were important to justify a 'carbon' project – but which nevertheless conveniently overlooked some key uncertainties and alternative perspectives (Leach and Scoones, 2013).

Demarcating the project boundary – a necessary first stage – was relatively straightforward since this was defined to coincide with the boundary of the reserve itself, as re-demarcated by 2011 to cover existing areas of 'intact' forest (OBF, 2012; Moninger, 2011). However, the requirement of a reference area of comparable vegetation outside the project boundary proved difficult to meet. The project boundary coincided with the area of currently so-called 'intact' forest with vegetation outside it in various states of settlement, farmland and bush-fallow regeneration. Thus, it was non-comparable. There was no way to surmount the difficulty and so the project remained VCS non-compliant in that respect.

Secondly, the project boundary was assumed to define ownership of the land, or at least the carbon within it. Again, in this case it appeared relatively straightforward to ascribe ownership to the government, in line with the formal view that the reserve is government land. This certainly made the project and ownership of its potential revenues more straightforward to administer – assuming all of it is government owned. Yet as expressed in the previous section, some community members contested this tenurial arrangement, claiming pre-colonially grounded land rights. Whether these will surface in claims to carbon rights and financial benefits when the succeeding and ongoing National REDD+ programme begins to generate revenues remains to be seen, but seems likely.[2]

To ensure eligibility, the project used the Food and Agriculture Organization's (FAO) definition of a forest, in the absence of a specific national or project definition. According to that definition, the entire WAPFoR constitutes a forest, and so is eligible for VCS crediting. But the definition excludes land under bush fallow cycles and village agroforestry management, such as exist today on the forest fringe and which were parts of the peninsula forests' history (Fairhead and Leach, 1998). Thus, definitions link carbon forestry exclusively with currently 'intact' high forest devoid of people, excluding any potential for generating carbon credits from currently lived-in landscapes.

Key to establishing the feasibility and justification for the carbon project were assessments of the baseline scenario in the absence of future project intervention. The OBF study compared satellite data from 2000 with that from 2006 and 2011, producing a set of land and forest change maps. From this they concluded 'quite some change (9 per cent) on the Western Area Peninsula, particularly around Freetown and Waterloo' (OBF, 2012, p.12). They then projected forward to 2031 by linking this historical baseline deforestation to the assumed 'major driver of deforestation which is urban expansion due to population growth and urbanization' by extrapolating apparent trends in population for 1985 and 2004 (OBF, 2012, p.4; Koroma *et al.*, 2006). Thus, they produced two alternative baseline deforestation scenarios, assuming relatively faster or slower deforestation, both suggesting a sufficient level of 'avoided deforestation' for the project to proceed. The OBF reported a mitigation potential on the condition that it was ground-truthed by forest inventory. This was undertaken in early 2013 by the University of Sierra Leone (USL), and the findings indicated that the WAPFoR had good carbon stocks (Karim *et al.*, 2013). By the end of the WAPFoR Project, the same consultants were busy trying to quantify potential emission reductions by assessing 'leakages' – assurance that if a carbon emitting activity is prevented in the project area, it is not simply displaced to nearby areas, resulting in no net decline in emissions overall – and identifying the threats (internal and external) that may affect the permanency of the action.

Following these activities, the project brochure publicized that: 'Based on our projections, the deforestation is likely to continue in the future by 280 to 440 hectares per year. Thus, the project is capable to achieve 80,000 to 130,000 tCO2 avoided carbon emissions per year from 2012 to 2031' (WHH, 2014, p.2). Linking this to even conservative estimates of possible carbon prices, the project was said to offer 'a promising investment opportunity' (WHH, 2014, p.1).

However, these calculations overlooked some key social and ecological dynamics in the landscape. Not acknowledged in OBF's analysis was forest expansion in some rural areas south of the reserve; a process perhaps linked to villagers moving out of relatively unprofitable farming in favour of fishing using improved energy-saving processing technologies and more so ecotourism. Yet attention to such trends would undermine the picture of rapid deforestation outside the project area, threatening the reserve, which is so central to the justification of the carbon project. Assuming that future trends can be inferred from past practices was also problematic given that – as the history of the WAPFoR has shown – forest landscape change and its drivers are not necessarily linear. As we have seen, land-use trends on the peninsula have been affected by the coming and going of foreign investments and industries; by war and migration; by foreign building projects; and by the complex political dynamics of land investment and settlement. Assuming steady linear trends in such a context is misguided.

Thus, the overall narrative constructed through the assessment was of an intact forest under steadily increasing threat from urban encroachment and local livelihood activities. Conservation capacity was weak, and required reinforcement (Winnebah, 2010); only the envisaged project would be able to provide this, in order to arrest deforestation that would happen otherwise.[3] This is, of course, a familiar narrative of the sort that has long justified conservation in the WAPFoR and, indeed, as we have seen, core project objectives and approaches – strengthening forest protection, and supporting alternative livelihood activities – remain largely unchanged. However, the terminology of 'business as usual scenarios', additionality, permanence and leakage serves to give a new gloss and authority to this older storyline. Linear deforestation narratives at the hands of local populations are repeated and strengthened with new force. And old conservation approaches are newly legitimated and strengthened, now as means to meet the demands of and generate profits within the new carbon economy. Maintaining a forest fortress is even more vital when it houses a potentially profitable carbon stock.

Community understandings, livelihood impacts

From the viewpoint of the project then, the WAPFoR's turn to carbon involves a mixture of continuity and change. But how is the project being understood, interpreted and experienced by communities living around the reserve? How is the project impacting on people's livelihoods? What difference does the turn to carbon make to them?

People living in the areas surrounding the WAPFoR make use of a wide range of forest-related resources in their everyday activities and livelihood strategies. These include land for farming; bees for honey; plants of medicinal values; branches for household fuelwood use, charcoal burning, smoking fish and baking bread and cakes for sale; timber for house and boat construction; stones for building construction; and stream and river water. Resource use is strongly differentiated by gender. For instance, women use trees (mainly deadwood) for fuelwood and charcoal burning, and land for horticulture, while men are usually responsible for building and boat construction. A few of these resource-using activities depend on land and trees within the forest reserve – including access to building and boat-building timber, and the sale of land to immigrants. However, most other resources can be more easily and conveniently obtained outside the reserve and buffer zone, within or near household compounds, governed by local social relations and access arrangements.

Views and experiences of the WAPFoR forest conservation project in general are varied, depending on people's social and livelihood positions. Some are appreciative. Thus, many community members value the re-demarcation of the reserve boundary in helping to reduce what they see as illegal sale of land to immigrants – its 'grabbing'. Youth and elders alike value enhanced forest protection in helping to control unwanted power saw operators, who come into the forest from urban areas and the provinces – although some, who have helped facilitate such exploitation and profited from it, take a different view. Women emphasize that they are happy to see a reduction in tree cutting as this reduces damage to houses by strong winds from the mountains, and improves the availability of adequate and safe drinking water. Men emphasize appreciation for new employment opportunities as forest guards, and in ecotourism – although these remain limited to a few.

However, others have been anxious about the enforcement of forest conservation, seeing this as restricting their use of valued timber forest products (e.g. for building houses and fishing boats) and their use of agricultural land. Immigrants – who make most use of activities within the forest reserve – indicated their unhappiness at being asked to move out of it (Winnebah, 2008). In efforts to secure local acceptance and participation, ENFORAC held public meetings in 44 of 52 target peri-urban and rural communities, during which formal Memoranda of Understanding (MoUs) on the roles, rights and duties of communities in environmental protection were to be signed in the presence and with the support of MAFFS and the EPA-SL. While many community authorities did sign, others (including the leaders of Adonkia, Ogoo Farm, Hamilton, Mambo, Sussex, Baw Baw, No 2 River and John Obey) refused because of apprehensions that communities could not adhere to these commitments without undue costs to their livelihoods (Winnebah, 2006, 2008, 2011). Some also expressed unwillingness to take responsibility for conservation, which they felt should be the government's responsibility. While there has been little overt resistance to forest protection activities, a few forest guards have experienced attacks – although whether from outside forest exploiters, impoverished immigrants or long-term community members remains unclear.

The project reported having trained EMCs comprising of the village head (wo) man, the secretary to the head (wo)man, the women's leader, the youth representative, a social worker, an animator and a community elder in all targeted communities, in natural resource management and the involvement of key forest stakeholders. EMCs were in turn supposed to coordinate forest conservation and management of environmental and livelihood activities on behalf of the communities. Alternative livelihood activities supported by the WAPFoR project and partners include motorbikes or 'Okada' supplied as alternative livelihoods for wood cutters and youths; small-scale irrigation for intensified agriculture; beekeeping; and the dissemination of improved fish-smoking ovens aimed at increasing women's incomes, while reducing pressure on fuelwood use from the forest. The project supported training workshops in micro-finance and business development in six communities, focusing on women and offering them micro credit to start small businesses. As a result of such activities, EMC observers state that there are now fewer people whose livelihoods depend on forest resources. However, many community members commented that the livelihood projects are patchy in spread, with only a few communities able to take part in each activity. In many cases, communities reported having been promised benefits that did not materialize. Some expressed their 'disgust' with the WAPFoR project because of this. Where people (mostly recent immigrants) have been dependent on resources from within the reserve, alternatives have not provided an adequate substitute. Following initial investment, resources for maintenance have often not been forthcoming so that infrastructure and technologies have fallen into disrepair.

The focus on alternative livelihoods as a way of reducing deforestation is problematic in other ways. It is not clear that the livelihoods supposed to be rendered more sustainable in this way were ever dependent on high forest within the reserve in the first place. Fuelwood and sites for horticulture are more conveniently obtained in areas close to villages, held within a bush-fallow-farm landscape. The provision of alternative rural livelihoods has also proved an inadequate means to address the issue of 'land grabbing'. As we have seen, this – the major driver of what recent deforestation has taken place in and around the reserve – is largely driven by a dynamic involving state officials, Freetown elites and immigrants, only sometimes in alliance with local community members. As senior Ministerial figures recognize, redressing this dynamic will require institutional and political action at the highest levels, including the regularization of land and municipal laws and responsibilities. Some relevant institutional reforms are underway as part of a broader process of redefining Sierra Leone's land policy. However, in the meantime, the assumption that this key deforestation dynamic is appropriately addressed by educating and providing alternative livelihoods to rural community members seems misguided at best.

The project also engaged in many environmental education and sensitization efforts, ranging from organized visits to the reserve's boundary with forest guards, to community outreach meetings, film shows, radio discussions, education conducted through the EMCs and the training of school teachers including school and youth clubs. These efforts have largely followed the long-established project narrative focusing on the importance of forest conservation to protect water supplies and biodiversity (Winnebah, 2006, 2011), and the need for community co-operation to support reserve boundary protection (Winnebah, 2008) and halt the activities of 'woodcutters, stone miners and charcoal burners'. These activities, their impacts and community members' perceptions of them are typical of a conservation-with-development project, in which education and livelihood activities are used – with limited success – to secure local acceptance of an essentially fortress-style set of forest protection activities.

So what difference does the proposed turn to carbon financing make? Not much from a community angle. The activities continue largely as before, although the project's managers increasingly see them as geared to protecting a carbon stock. Indeed, there is limited local awareness of the project's plans around carbon. Most community members interviewed reported that they had no idea of future WAPFoR plans. Even NGO partners claimed that planning meetings were seldom held, because much of the planning, financial disbursement and implementation were monopolized by the WHH Project Manager. This reflects the complexities of scoping and developing carbon projects, in which negotiations with consultants – foreign and local technical experts – have taken precedence over discussions with local partners and communities – for whom the technicalities are deemed too complex.

Nonetheless, rumours, whether picked up from project outreach workers, media or social networks, have filtered out about future carbon developments. Terms such as climate change, carbon and 'REDD' have become common currency amongst people living around the WAPFoR reserve. People have developed their own ways of interpreting these difficult-to-fathom concepts, their relationships with forests and the presumed actions of outsiders in relation to them. Local understandings draw on both everyday and historically-embedded rationalities and experiences.

When asked specifically about climate change and its causes, many community members and forest guards alike said they 'did not know the meaning'. Young men in Baoma said, 'It simply means the changing of the weather condition' (focus group discussion, 19 September 2013). Others had more elaborate local theories. For example, men at Bonga Wharf said, 'yes, when you burn the bush, the smoke that goes up is climate change. The stones that expose give up heat [...] they cut the trees that help to keep the environment cool' (focus group discussion, 10 September 2013). Some did link climate change with industrial emissions, as put by a group of young men: 'It is caused by deforestation and air pollution' (group discussion, 19 September 2013), while CBO members and officials at Bathurst said, 'it is caused by waste products from extractive industries'.

Villagers who perceived a relationship between forests and climate generally constructed this in local terms – unsurprising, given the long history of interventions in the WAPFoR area aimed at forest protection for local water supplies. Thus, community leaders at Baoma said, 'forests regulate climate through the water cycle', while members of the EMC at Kossoh Town said, 'when forest is lost, it cannot rain on time. We can get extinction of animals'. CBO officials at No 2 River said, 'forests catch rain and form clouds'. Equally unsurprisingly, forestry and environmental protection officials elaborated theories more in line with the international scientific discourses to which they have been exposed. Thus, the SLP's OSD personnel at Mile 13 said, 'forests absorb carbon dioxide from advanced countries leading to increasing rainfall'. An EPA-SL official said, 'in the absence of forest, there will be increasing carbon dioxide leading to increasing warming that eventually results in drought and dry spells'.

When asked what they thought carbon was, responses were imaginative and creative – yet again, logical in terms of local experience. Community leaders variously described carbon as 'smoke from burning wood', 'diffused gas caused by combustion or charcoal burning' and 'smoke from vehicles'; while a forest guard at No 2 River erroneously said, 'from trees. Through photosynthesis plants give out carbon dioxide'. Women at Brigitte Town said, 'Carbon is the smoke that you get when you smoke fish. It is also the smoke you get when you burn wood', while those at Deep Eye Water said, '[you] use it to write a duplicate receipt'. Young men variously imagined carbon as 'smoke from trees in the forest', a 'black substance from charcoal and smoke from factory', and 'gas from the soil and forest'. A group of women said that carbon was 'the mist you see above the forest in the morning'. Some expressed more elaborate narratives around the link between carbon and forests, reflecting a mixture of scientific concepts and local experience. Thus, community leaders at Bathurst said that 'forest help to condense carbon dioxide', while immigrants at Baoma said that 'smoke is absorbed by trees'.

Most community members and forest guards claim to have little, if any, knowledge of carbon market schemes and initiatives, either in the WAPFoR or more broadly. Young men at Kossoh Town and Baoma had heard that carbon could be sold, but were puzzled and intrigued, asking 'how is it done?' and 'who sells it?' A few had heard of 'REDD' but did not know any details. However others, including members of EMCs who had had more exposure and opportunity to reflect on possible meanings, offered elaborate theories about REDD and other carbon market initiatives. Thus, a man at Baoma said that REDD 'is for communities with reserve forest where money is poured into its account for other development activities'. In a group discussion with EMC members at Brigitte Town (23 October 2013), it was said that 'WAPFoR said to come out from the bush, because they need carbon in other countries, so they need to buy the bush'; that 'We think it is best to let the forest grow big, and build up much carbon [...] WAPFoR will use it to sell, and get money back for the country'; and that 'The EU will give a guarantee, that if they do industry in Sierra Leone, and Sierra Leone guarantees carbon, then the EU will invest more'. Forest guards in the Kent Village Junction area (group discussions on 23 October 2013) described how 'We have heard that European countries need the carbon. They will come and buy it. Then the government will have money to spend on employment'. They went on to say, 'we have to protect the forest well so that there will be plenty of carbon when they come'. Perhaps most tellingly, when the research team arrived for one EMC group discussion at Briggitte Town on 23 October 2013, informants asked 'Are you Reg (aka REDD) who has come to buy our carbon?'

In such statements, carbon is imagined as a saleable commodity to be extracted by foreigners and exported. In local terms, it is thus seen as rather like foreign mineral or timber extraction – repeated themes in the history of intervention in Sierra Leone's forest landscapes (Richards, 1996). Notably, though, this is also a rather accurate description of the global carbon economy, and its engagement with African forests and people.

Carbon in the WAPFoR area is thus imagined centrally as a source of future money – by both the project developers and local people, albeit in different ways. But, how will these promised future financial benefits be shared? At the moment, it seems that communities do not imagine a significant share of carbon profit. When asked explicitly how they think benefits should be shared, most villagers state that the greatest share should finance community infrastructure and social development. Yet they have had little opportunity to consider this in any detail, while experience with minerals has taught them that the bulk of the profits from extractive resources usually go to foreigners, or to the government. Indeed, it seems that the project developers have a similar view. The WAPFoR Project Manager asserts that the WAPFoR is 'state land and not community as in the hinterland of the country [...] and requires no benefit sharing of carbon money' (interview, 3 and 24 October 2013). In these terms, formal tenure arrangements give communities no land or carbon rights within the reserve, and hence no rights to benefits. Instead, it is envisaged that community benefits are tokenistic, and come through alternative livelihood activities and local income generating and employment projects, as described above. Yet, as we have seen, these benefits are meagre for most. Although it might be expected that under a carbon project they could be increased, the project developers are clear that strengthening forest protection with more (armed) guards and fences is the priority. If low carbon prices means that funds are tight, then even small community benefits will have to be sacrificed.

Yet, as we have seen, government ownership of the reserve land – and thus its carbon – is contested by those who claim ancestral rights based on pre-colonial occupation and farming. If carbon profits really begin to flow, one can envisage such claims resurfacing, and serving to fuel conflict between local people and the government over whose carbon and whose money is at stake.

Conclusions

The WAPFoR case thus exemplifies powerfully the ways in which new carbon interventions geared to the global carbon economy become layered into landscape and intervention histories. Over the course of a century, Sierra Leone's WAPFoR has been successively treated as: farming and settlement sites for pre-colonial Temne and Sherbro populations; sources of European shipbuilding timber; protectors of watersheds and water supplies; havens of biodiversity, rare animals and birds; aesthetic attractions for urban dwellers and tourists; sources of land for urban settlers and foreign elite investments; places of timber and non-timber resources for local livelihoods; safe havens for internally displaced people and refugees; launching grounds for regime change; and protectors of local climate. Now, on top of all this, they are seen as carbon stocks: sources of commodities for eventual sale.

This series of interventions has involved successive landscape transformations – ecologically, socially, tenurially, and discursively. Each is influenced by previous interventions: while today's WAPFoR is idealized as a place of high, dense trees for carbon, its diverse fauna still inhabits the area and the kola and fruit trees that marked pre-colonial settlements have left their vegetational legacies. While carbon project plans and methodologies assume straightforward 'fortress' conservation of a government reserve, older tenurial claims continue to resurface. And local forest personnel and community members alike imagine 'carbon' and the prospects of carbon market initiatives and benefits in ways that evoke both long-established ideas and interventions around watershed protection, and Sierra Leone's history of foreign resource extraction.

The carbon project's narratives gloss over much of this complex landscape history. Project documents and methodologies instead paint a picture of ongoing linear deforestation due to urban expansion and local livelihood activities placing increasing pressure on an 'intact' forest reserve, extending from the past into the future. Such a narrative is necessary to justify the carbon project, but it is a poor representation of real social, political and ecological dynamics. And it supports a strengthened, fortress approach to forest protection that excludes local people from any real stake in managing the carbon project or sharing in its future benefits.

While such a gulf persists between project narratives and landscape history as locally understood and experienced, WAPFoR's claims to be developing a 'pro-poor' and 'participatory' carbon project will be difficult to realize in practice. The project has, in principle, been engulfed into the recent ongoing national REDD+ process that is currently building capacity and preparing readiness for the sale of carbon stocks in the country's reserves as a whole. As it develops further in that context, serious attention to community members' rights, claims and livelihood needs will be essential. Urban expansion pressures will need to be addressed as the political and institutional problem they are, not imagined as solvable through more forest guards and alternative livelihood activities. Efforts will need to be made to communicate, translate and reconcile local imaginaries of carbon, climate and the carbon economy, with project plans and narratives. Otherwise, and despite the best intentions of project developers, a scheme intended to promote sustainable forest use for global and local interests will end up intensifying conflicts between these, ultimately undermining the project itself.

Notes

1   According to an interview with the WAPFoR Project Manager, an additional 26 villages 'received various intervention services from the project on "compassionate" grounds given that they were not budgeted for' in the approved grant application (interview, WAPFoR Project, Freetown, 24 October 2013).

2   Invitees in a follow-on policy-project-community dialogue agreed that meaningful consultations with all stakeholders at the beginning is paramount to establish clarity with respect to ownership rights, demarcations and enforcement in a policy document and programming activities. One headman observed that 'although government is said to own the reserve and we cannot challenge that [...] anything short of that would result in us taking up machetes and sticks to defend our rights' (meeting, Waterloo Town, 26 September 2014). The Mayor of Freetown stressed that, 'as far as the land policy issue is concerned, there is bound to be several conflicting areas in this country in the not too distant future' (meeting, Freetown, 24 September 2014).

3   The ongoing REDD+ programme will 'engage the services of the DFOs and Forest Guards in all the districts. These DFOs and Forest Guards will be needed to be engage in constant field monitoring and to identify the drivers of deforestation and degradation within their area of operation. This activity will be carried out by the DFOs and Forest Guards through the means of community outreach and field monitoring processes' (RoSL, MAFFS and EU, 2014, p.19).

REFERENCES

Adaptive Research Planning Team (ARPT) (1984) *Lusaka Province Zoning Report*, field work by T. G. Maynard, C. E. A. Masi, A. J. Sutherland, S. A. Kean and C. A. Njobvu, Adaptive Research Planning Team.

Agrawal, A., Lemos, M. C., Orlove, B. and Ribot, J. (2012) 'Cool heads for a hot world – Social sciences under a changing sky', *Global Environmental Change*, vol 22, no 2, pp.329–331.

Ainembabazi, J. H. (2010) *Landlessness Within the Vicious Cycle of Poverty in Ugandan Rural Farms Households: Why and How is it born?*, Economic Policy Research Centre, Kampala, http://dspace.cigilibrary.org/jspui/handle/123456789/29982, accessed 14 December 2013.

Akitanda, P. (2002) *South Kilimanjaro Catchment Forest Project*, Moshi, Kilimanjaro.

Allen, J. and Cochrane, A. (2010) Assemblages of State Power: Topological Shifts in the Organization of Government and Politics, *Antipode*, vol 42, no 5, pp.1071–1089.

Amanor, K. S. (1994) *The New Frontier: Farmers' Response to Land Degradation. A West African Study*, Zed Books, London.

Angelsen, A. (ed) (2008) *Moving Ahead with REDD: Issues, Options and Implications*, Center for International Forestry Research (CIFOR) Bogor, Indonesia.

Angelsen, A. (2013) 'REDD+ as performance-based aid: General lessons and bilateral agreements of Norway', UNU-WIDER Working Paper, no 135, UNU-WIDER, Helsinki, Finland.

Angelsen, A., Brockhaus, M., Kanninen, M., Sills, E., Sunderlin, W. D. and Wertz-Kanounnikoff, S. (eds) (2009) *Realising REDD+ National Strategy and Policy Options*, Center for International Forestry Research (CIFOR) Bogor, Indonesia.

Angelsen, A., Brockhaus, M., Sunderlin, W. D. and Verchot, L. V. (eds) (2012) *Analysing REDD+: Challenges and Choices*, Center for International Forestry Research (CIFOR) Bogor, Indonesia.

Appadurai, A. (1986) *The Social Life of Things*, Cambridge University Press, Cambridge.

Arhin, A. A. (2014) 'Safeguards and dangerguards: A framework for unpacking the black box of safeguards for REDD+', *Forest Policy and Economics*, vol 45, pp.24–31.

Ascui, F. and Lovell, H. (2011) 'As frames collide: Making sense of carbon accounting', *Accounting, Auditing and Accountability Journal*, vol 24, no 8, pp.978–999.

Atela, J. O. (2013) 'Governing REDD+: Global framings versus practical evidence from the Kasigau Corridor REDD+ Project, Kenya', *STEPS Working Paper*, no 55, STEPS Centre, Brighton, http://steps-centre.org/wp-content/uploads/Governing-REDD+.pdf, accessed 9 October 2014.

Atela, J. O., Quinn, C. H. and Minang, P. A. (2014a) 'Are REDD projects pro-poor in their spatial targeting? Evidence from Kenya', *Applied Geography*, vol 52, pp.14–24.

Atela J. O., Quinn C. H., Minang, P. and Duguma, L. (2014b) 'Nesting REDD+ into integrated conservation and development projects: What empirical lessons can be drawn?' *Centre for Climate Change Economics and Policy Working Paper*, no 182, www.see.leeds.ac.uk/fileadmin/Documents/research/sri/workingpapers/SRIPs-68.pdf, accessed 9 October 2014.

Auty, R. M. (1993) *Sustaining Development in Mineral Economies: The Resource Curse Thesis*, Routledge, London.

Bakarr, M. I., Bailey, B., Omland, M., Myers, N., Hannah, L., Mittermeier, C. G and Mittermeier, R. A. (1999) 'Guinean forests', in R. A. Mittermeier, N. Myers, P. R Gil and C. G. Mittermeier (eds) *Hotspots: Earth's Biologically Richest and Most Endangered Terrestrial Ecoregions*, Toppan Printing Company, Japan, pp.239–253.

Balée, W. (2006) 'The research program of historical ecology', *Annual Review of Anthropology*, vol 35, pp.75–98.

Bateman, I. J., Mace, G. M., Fezzi, C., Atkinson, G. and Turner, K. (2011) 'Economic analysis for ecosystem service assessments', *Environmental and Resource Economics*, vol 48, no 2, pp.177–218.

Bayart, J. F. (1993) *The State in Africa: The Politics of the Belly*, Longman, London and New York.

Benessaiah, K. (2012) 'Carbon and livelihoods in Post-Kyoto: Assessing voluntary carbon markets', *Ecological Economics*, vol 77, pp.1–6.

Berry, S. (1989) 'Social institutions and access to resources', *Africa*, vol 59, no 1, pp.41–55.

Berry, S. (1993) *No Condition is Permanent: The Social Dynamics of Agrarian Change in sub-Saharan Africa*, University of Wisconsin Press, Wisconsin.

Berry, S. (2002) 'Debating the land question in Africa', *Comparative Studies in Society and History*, vol 44, no 4, pp.638–668.

Biddulph, R. (2011) 'Is the Geographies of Evasion hypothesis useful for explaining and predicting the fate of external interventions? The case of REDD in Cambodia', paper presented at the conference 'Globalization and Development: Rethinking interventions and governance', University of Gothenburg, 22–23 November.

BioCarbon Partners (BCP) (2013) *Lower Zambezi REDD+ Project, Rufunsa District, Zamiba*, Project Design Document, BioCarbon Partners, Lusaka, https://s3.amazonaws.com/CCBA/Projects/Lower_Zambezi_REDD%2B_Project/Validation/BCP_LowerZambezi_REDD+Project_PDD_CCB_21Jun_2013.pdf, accessed 25 October 2014.

BioCarbon Partners (BCP) (no date) *Community Projects Outline*, unpublished.

BioCarbon Partners (BCP) Trust (2013) Newsletter, 2, www.biocarbonpartners.com, accessed 10 July 2014.

Bird, C. and Metcalfe, S. (1996) 'Two views from CAMPFIRE in Zimbabwe's Hurungwe District: Training and motivation. Who benefits and who doesn't?', *IIED Wildlife and Development* Series, no 5, IIED, London.

Bird, N. (2012) 'Approved methodology for the Adoption of Sustainable Agricultural Land Management', Verified Carbon Standard Methodology, VM0017, Version 1.0 Sectoral Scope 14.

Blaikie, P. and Brookfield, H. (1987) *Land Degradation and Society*, Methuen, London.

Blom, B., Sunderland, T. and Murdiyarso, D. (2010) 'Getting REDD to work locally: Lessons learned from integrated conservation and development projects', *Environmental Science & Policy*, vol 13, no 2, pp.164–172.

Böhm, S. and Dabhi, S. (eds) (2009) *Upsetting the Offset: The Political Economy of Carbon Markets*, MayFly Books, London.

Bond, I., Chambwera, M., Jones, B., Chundama, M. and Nhantumbo, I. (2011) *REDD+ in Dryland Forests: Issues and Prospects for Pro-Poor REDD in the Miombo Woodlands of Southern Africa*, Natural Resources Issue 21, IIED, London.

Bond, I., Grieg-Gran, M., Wertz-Kanounnikoff, S., Hazlewood, P., Wunder, S. and Angelsen, A. (2009) *Incentives to Sustain Forest Ecosystem Services: A Review and Lessons for REDD*, IIED, London.

Boone, C. (2003) *Political Topographies of the African State: Territorial Authority and Institutional Choice*, Cambridge University Press, Cambridge.

Boone, C. (2013) *Property and Political Order in Africa: Land Rights and the Structure of Politics*, Cambridge University Press, New York.

Booth, D. (2011) 'Introduction: Working with the grain? The Africa power and politics programme', *IDS Bulletin*, vol 42, no 2, pp.1–10.

Boudron, F., Andersson, J. and Giller, K. (2011) 'Failing to yield? Ploughs, conservation agriculture and the problem of agricultural intensification: An example from the Zambezi Valley, Zimbabwe', *Journal of Development Studies*, vol 48, no 3, pp.393–412.

Boyd, E. (2002) 'The Noel Kempff project in Bolivia: Gender, power and decision-making in climate mitigation', *Gender and Development*, vol 10, no 2.

Boyd, E. and Goodman, M. K. (2011) 'The clean development mechanism as ethical development? Reconciling emissions trading and local development', *Journal of International Development*, vol 23, no 6, pp.836–854.

Boyd, E., Boykoff, M. and Newell, P. (2011) 'The "new" carbon economy: What's new?', *Antipode*, vol 43, no 3, pp.601–611.

Boyd, E., Corbera, E., Gutierrez, M. and Estrada, M. (2004) 'The politics of afforestation and reforestation activities at COP-9 and SB-20', *Tyndall Briefing Paper*, no 71, Tyndall Centre, Norwich.

Boyd, E., Gutierrez, M. and Chang, M. (2007) 'Small-scale forest carbon projects: Adapting CDM to low-income communities', *Global Environmental Change*, vol 17, no 2, pp.250–259.

Boyd, E., Hultman, N., Timmons Roberts, J., Corbera, E., Cole, J., Bozmoski, A., Ebeling, J., Tippman, R., Mann, P. and Brown, K. (2009) 'Reforming the CDM for sustainable development: Lessons learned and policy futures', *Environmental Science & Policy,* vol 12, no 7, pp.820–831.

Boyd, E., May, P., Chang, M. and Veiga, F. C. (2007) 'Exploring socioeconomic impacts of forest based mitigation projects: Lessons from Brazil and Bolivia', *Environmental Science & Policy*, vol 10, no 5, pp.419–433.

Brandon, K. and Wells, M. (2009) 'Lessons for REDD+ from protected areas and integrated conservation and development projects', in A. Angelsen (ed) *Realising REDD+National Strategy and Policy Options*, Center for International Forestry Research (CIFOR), Bogor, Indonesia.

Braun, B. and Castree, N. (eds) (1998) *Remaking Reality: Nature at the Millennium*, Routledge, London and New York.

Bridge, G. (2011) 'Resource geographies 1: Making carbon economies, old and new', *Progress in Human Geography*, vol 35, no 6, pp.820–834.

Brockington, D. (2002) *Fortress Conservation: The Preservation of the Mkomazi Game Reserve, Tanzania*, James Currey, Oxford.

Brockington, D. (2011) 'Ecosystem services and fictitious commodities', *Environmental Conservation*, vol 38, no 4, pp.367–369.

Brockington, D., Sachedina, H. and Scholfield, K. (2008) 'Preserving the new Tanzania: Conservation and land use change', *International Journal of African Historical Studies*, vol 41, no 3, pp.557–579.

Brown, D., Seymour, F. and Peskett, L. (2008) 'How do we achieve REDD co-benefits and avoid doing harm?', in A. Angelsen (ed) *Moving Ahead with REDD Issues, Options and Implications*, Centre for International Forestry Research (CIFOR), Bogor, Indonesia.

Brown, K. and Corbera, E. (2003) 'Exploring equity and sustainable development in the new carbon economy', *Climate Policy* vol 3, S1, pp.41–56.

Brown, K., Adger, W. N., Boyd, E., Corbera-Elizalde, E. and Shackley, S. (2004) *How do CDM Projects Contribute to Sustainable Development?*, Tyndall Centre for Climate Change Research, Norwich.

Brown, M. I. (2013) *Redeeming REDD: Policies, Incentives, and Social Feasibility in Avoided Deforestation*, Earthscan/Routledge, London and New York.

Bumpus, A. G. (2012) 'The matter of carbon: Understanding the materiality of tCO2e in carbon offsets', in P. Newell, M. Boykoff and E. Boyd (eds) *The New Carbon Economy: Constitution, Governance and Contestation*, John Wiley, Chichester.

Büscher, B. (2011) 'Nature on the move: The emergence and circulation of fictitious conservation and liquid nature', paper presented at the conference on 'NatureTM Inc? Questioning the market panacea in environmental policy and conservation', Institute of Social Studies, The Hague, 30 June–2 July.

Büscher, B. and Dressler, W. (2012) 'Commodity conservation: The restructuring of community conservation in South Africa and the Philippines', *Geoforum*, vol 43, no 3, pp.363–656.

Büscher, B., Sullivan, S., Neves, K., Igoe, J. and Brockington, D. (2012) 'Towards a synthesized critique of neoliberal biodiversity conservation', *Capitalism Nature Socialism*, vol 23, no 2, pp.4–30.

Çalışkan, K. and Callon, M. (2009) 'Economization, part 1: Shifting attention from the economy towards processes of economization', *Economy and Society*, vol 38, no 3, pp.369–98.

Çalışkan, K. and Callon, M. (2010) 'Economization, part 2: A research programme for the study of markets', *Economy and Society*, vol 39, no 1, pp.1–32.

Callon, M. (2009) 'Civilizing markets: Carbon trading between in vitro and in vivo experiments', *Accounting, Organizations and Society*, vol 34, pp.535–48.

Calmel, M., Martinet, A., Grondard, N., Dufour, N., Rageade, M. and Ferté-Devin, A. (2010) *REDD+ at Project Scale: Evaluation and Development Guide*, ONF International.

Castree, N. (2008a) 'Neoliberalising nature: The logics of deregulation and reregulation', *Environment and Planning A*, vol 40, no 1, pp.131–152.

Castree, N. (2008b) 'Neoliberalising nature: Processes, effects, and evaluations', *Environment and Planning A*, vol 40, no 1, pp.153–173.

CGIAR Research Program on Climate Change, Agriculture and Food Security (CCAFS) (2009) *Climate Change, Agriculture and Food Security, A CGIAR Challenge Program*, The Alliance of the CGIAR Centres and ESSP, Copenhagen, Denmark.

CGIAR Research Program on Climate Change, Agriculture and Food Security (CCAFS) (2013) 'Global Landscapes Forum: Shaping the climate agenda for forests and agriculture', http://ccafs.cgiar.org/events/16/nov/2013/global-landscapes-forum#.U6Muifk7vLk, accessed 15 July 2014.

CDI (2012) *Jungle Fever*, Climate and Development Initiatives, Kampala.

Center for International Forestry Research (CIFOR) (2014) 'Landscape approach dovetails with REDD', http://blog.cifor.org/20524/landscapes-approach-dovetails-with-redd-scientist-says#.U4MQj01OXIU, accessed 26 May 2014.

Chabal, P. and Daloz, J. P. (1999) *Africa Works: Disorder as Political Instrument* (African Issues), James Currey, Oxford.

Chanock, M. (1991) 'A peculiar sharpness: An essay on property in the history of customary law in colonial Africa', *Journal of African History*, vol 32, no 1, pp.65–88.

Chhatre, A., Lakhanpal, S., Larson, A. M., Nelson, F., Ojha, H. and Rao, J. (2012) 'Social safeguards and co-benefits in REDD+: A review of the adjacent possible', *Current Opinion in Environmental Sustainability*, vol 4, no 6, pp.654–660.

Chidumayo, E. N. (2001) *CHAPOSA. Charcoal Potential in Southern Africa. Final Report for Zambia*, University of Zambia and Stockholm Environment Institute, Lusaka.

Chimhowu, A. (2002) 'Extending the grain basket to the margins: Spontaneous land resettlement and livelihoods in Zimbabwe', *Journal of Southern African Studies*, vol 38, no 3, pp.551–573.

Chimhowu, A. (2003) 'Land resettlement and livelihoods in Zimbabwe', unpublished thesis, Institute for Development Policy and Management, University of Manchester.

Chimhowu, A. and Hulme, D. (2006) 'Livelihood dynamics in planned and spontaneous resettlement in Zimbabwe: Converging and vulnerable' *World Development*, vol 34, no 4, pp.728–750.

Chivuraise, C. (2013) *Economics of Smallholder Tobacco Production and Implication of Tobacco Growing on Deforestation in Hurungwe District of Zimbabwe*, MSc Thesis, Department of Agricultural Economic and Extension, University of Zimbabwe.

Chouin, G. (2009) 'Forests of power and memory: An Archaeology of sacred groves in the Eguafo Polity, Southern Ghana (c. 500–1900 A.D.)', PhD Dissertation, Anthropology, Syracuse University.

Clean Development Mechanism (CDM) (2013) http://cdm.unfccc.int/methodologies/SSCmethodologies/SSCAR/approved, accessed 15 January 2013.

Climate Community and Biodiversity Alliance (CCBA) (2010) *REDD+ Social and Environmental Standards*, CCBA, Arlington.

Colin, J. P. and Woodhouse, P. (2010) 'Introduction: Interpreting land markets in Africa', *Africa: The Journal of the International African Institute*, vol 80, no 1, pp.1–13.

Community Management of Protected Areas Conservation Trust (COMPACT) (n.d) 'Community management of protected areas project description document', Moshi, Tanzania, accessed from COMPACT office, April 2004.

Conservation International (2009) *Biodiversity Hotspots*, www.biodiversityhotspots.org/Pages/default.aspx, accessed 16 December 2012.

Corbera, E. and Brown, K. (2008) 'Building institutions to trade ecosystem services: Marketing forest carbon in Mexico', *World Development*, vol 36, no 10, pp.1956–1979.

Corbera, E. and Schroeder, H. (2010) 'Governing and implementing REDD', *Environmental Science and Policy*, vol 14, no 2, pp.89–99.

Corbera, E., Brown, K. and Adger, W. N. (2007) 'The equity and legitimacy of markets for ecosystem services', *Development and Change*, vol 38, no 4, pp.587–613.

Corson, C. and MacDonald, K. I. (2012) 'Enclosing the global commons: The convention on biological diversity and green grabbing', *Journal of Peasant Studies*, vol 39, no 2, pp.263–283.

Cosgrove, D. (1984) *Social Formation and Symbolic Landscape*, University of Wisconsin Press, Madison, Wisconsin.

Cotula, L. and Mayers, J. (2009) 'Tenure in REDD: Start-point or afterthought?' IIED Natural Resource Issues 16, IIED, London.

Crumley, C. L. (ed) (1994) *Historical Ecology. Cultural Knowledge and Changing Landscapes*, School of American Research Press, Santa Fe NM.

Das, V. and Poole, D. (eds) (2004) *Anthropology in the Margins of the State*, School of American Research Press, Santa Fe NM.

Davies, V. A. B. (2012) 'The political economy of government revenues in post-conflict resource-rich Africa: Liberia and Sierra Leone', *National Bureau of Economic Research Working Paper*, no 18539, Cambridge MA.

de Alcántara, C. H. (ed) (1993) *Real Markets: Social and Political Issues of Food Policy Reform*, Frank Cass, London.

Demeritt, D. (1994) 'The nature of metaphors in cultural-geography and environmental history', *Progress in Human Geography*, vol 18, no 2, pp.163–185.

Di Gregorio, M., Brockhaus, M., Cronin, T., Muharrom, E., Santoso, L., Mardiah, S. and Büdenbender, M. (2013) 'Equity and REDD+ in the media: A comparative analysis of policy discourses', *Ecology and Society*, vol 18, no 2, p.39.

Diaz, D., Hamilton, K. and Johnson, E. (2011) *State of the Forest Carbon Markets 2011: From Canopy to Currency*, Ecosystems Market Place/Forest Trends, Washington DC.

Djankov, S., Montalvo, J. G. and Reynal-Querol, M. (2008) 'The curse of aid', *Journal of Economic Growth*, vol 13, no 3, pp.169–194.

Dokken, T., Caplow, S., Angelsen, A. and Sunderlin, W. D. (2014) 'Tenure issues in REDD+ pilot project sites in Tanzania', *Forests*, vol 5, no 2, pp.234–255.

Dressler, W., Büscher, B., Schoon, M., Brockington, D. A. N., Hayes, T., Kull, C. A., McCarthy, J. and Shrestha, K. (2010) 'From hope to crisis and back again? A critical history of the global CBNRM narrative', *Environmental Conservation*, vol 37, no 1, pp.5–15.

Duncan, J. and Ley, D. (eds) (1993) *Place/Culture/Representation*, Routledge, London and New York.

Dzingirai, V. (2003) 'The new scramble for the African countryside', *Development and Change*, vol 34, no 2, pp.243–264.

Dzingirai, V., Andersson, J., Bourdon, F., Milgroom, J., Murungweni and Poshiwa, X. (2013) 'On the edge of the state and economy', in J. Andersson, M. DeGarinne-Wichatitsky, D. M. Cumming, V. Dzingirai and K. Giller (eds) *Transfrontier Conservation Areas: People Living on the Edge*, Routledge, London.

Ecofys (2014) *Mapping Carbon Pricing Initiatives: Developments and Prospects 2013*, World Bank, Washington DC.

Eklof, G. (2012) *REDD Plus or REDD 'Light'?: Biodiversity, Communities and Forest Carbon Certification*, Swedish Society for Nature Conservation, Stockholm.

Engel, S., Pagiola, S. and Wunder, S. (2008) 'Designing payments for environmental services in theory and practice: An overview of the issues', *Ecological Economics*, vol 65, no 4, pp.663–674.

Environment Africa (2011) *Kariba REDD Baseline Report*, Environment Africa, Harare.

Erickson, C. L. and Balée, W. (2006) 'The Historical Ecology of a Complex Landscape in Bolivia', in W. Balée and C. L. Erickson (eds) *Time and Complexity in Historical Ecology*, Columbia University Press, New York, pp.187–233.

EuropeAid/126201/C/ACT/Multi (2008) 'Environment and sustainable management of natural resources, including energy', Grant Application Form, Restricted Call for Proposals 2007–2008, Budget line 21 04 01 European Commission, 21 February.

Fairhead, J. and Leach, M. (1996) *Misreading the African Landscape: Society and Ecology in a Forest-Savanna Mosaic*, Cambridge University Press, Cambridge.

Fairhead, J. and Leach, M. (1998) *Reframing Deforestation: Global Analyses and Local Realities: Studies in West Africa*, Routledge, London.

Fairhead, J., Leach, M. and Scoones, I. (2012) 'Green grabbing: A new appropriation of nature?', *Journal of Peasant Studies*, vol 39, no 2, pp.237–261.

Fanthorpe, R. and Gabelle, C. (2013) *Political Economy of Extractives Governance in Sierra Leone*, report to the International Bank for Reconstruction and Development/The World Bank, Washington DC.

Fernandes, E., Oktingati, A. and Maghembe, J. (1984) 'The Chagga homegardens: A multistoried agroforestry cropping system on Mt. Kilimanjaro (Northern Tanzania)', *Agroforestry Systems*, vol 2, no 2, pp.73–86.

Fleming, G. (1966) 'Authority, efficiency, and role stress: Problems in the development of East African bureaucracies', *Administrative Science Quarterly*, vol 11, pp.386–404.

Fletcher, R. (2010) 'Neoliberal environmentality: Towards a poststructural political ecology of the conservation debate', *Conservation and Society*, vol 8, no 3, pp.171–181.

Fong Cisneros, J. A. (2012) *Forest Carbon Projects in Africa: A Mapping Study*, STEPS Centre Report, STEPS Centre, Brighton.

Food and Agriculture Organization of the United Nations (FAO) (2011) *Gender Differences in Assets*, ESA Working Paper 11–12, www.fao.org/docrep/013/am317e/am317e00.pdf, accessed 13 June 2012.

Forest Carbon Partnership Facility (FCPF) (2014) 'REDD+ Country Participants', https://www.forestcarbonpartnership.org/redd-country-participants, accessed 30 March 2014.

Forestry Division (2013) *Western Area Peninsula National Park: Management Plan 2014–2017*, Ministry of Agriculture, Forestry and Food Security, Sierra Leone.

Forsyth, T. (2003) *Critical Political Ecology: The Politics of Environmental Science*, Routledge, London.

Fortman, L. (1985) 'The tree tenure factor in agroforestry with particular reference to Africa', *Agroforestry Systems*, vol 2, no 4, pp.229–251.

Gallemore, C. and Munroe, D. K. (2013) 'Centralization in the global avoided deforestation collaboration network', *Global Environmental Change*, vol 23, no 5, pp.1199–1210.

Global Environmental Facility (GEF) (2010) 'Reducing land degradation on the highlands of Kilimanjaro region', project proposal, Washington DC.

Gockowski, J. and Sonwa, D. (2010) 'Cocoa intensification scenarios and their predicted impact on CO_2 emissions, biodiversity conservation, and rural livelihoods in the Guinea rain forest of West Africa', *Environmental Management*, vol 48, no 2, pp.307–21.

Gole, T. W. (2011) 'Carbon finance: Emerging opportunities for biosphere reserves in Africa', Environment and Coffee Forest Forum (ECFF), presentation on 'Managing Challenges of Biosphere Reserves in Africa', 27 June–2 July, Dresden, Germany, www.bfn.de/fileadmin/MDB/documents/themen/internationalernaturschutz/2011-AfriBR-03-Gole_Ethiopia.pdf, accessed 29 September 2014.

Goodman, M. K. and Boyd, E. (2011) 'A social life for carbon? Commodification, markets and care', *The Geographical Journal*, vol 177, no 2, pp.102–109.

Government of Ghana (2003) *National Development Planning Commission. Ghana Poverty Reduction Strategy: An Agenda for Growth and Prosperity: 2003–2005 [GPRS I]*, Government of Ghana, Accra.

Government of Kenya (1968) 'Land (group representation) Act', GoK, Nairobi, Kenya.

Government of Kenya (1999) The Kenya National Environmental Management and Coordination Act (EMCA, 1999), Government Printer, Nairobi, Kenya.

Government of Kenya (2005) 'Forests Act', GoK, Nairobi, Kenya.

Government of Kenya (2010a) 'Revised REDD+ Readiness Preparation Proposal', GoK, Nairobi, Kenya.

Government of Kenya (2010b) 'The Constitution of Kenya', GoK, Nairobi, Kenya.

Government of Kenya (2012) 'National Climate Change Action Plan', GoK, Nairobi, Kenya.

Government of the Republic of Zambia (GRZ) (1984) *Towards Self Sufficiency. A Blueprint for Agricultural Development in Lusaka Province*, Provincial Planning Unit, Lusaka.

Government of the Republic of Zambia (GRZ) (2009a) *Draft Forest Policy*, Forestry Department, Ministry of Tourism, Environment and Natural Resources, Lusaka.

Government of the Republic of Zambia (GRZ) (2009b) *Decentralization Implementation Plan 2009–2013*, Decentralization Secretariat, Ministry of Local Government and Housing, Lusaka.

Government of Zimbabwe (2013) *Zimbabwe National Climate Change Response Strategy*, Harare, Zimbabwe.

Grainger, M. and Geary, K. (2011) *The New Forests Company and its Uganda Plantations*, Oxfam International, Washington DC.

Greenberg, J. H. (1966) *The Languages of Africa*, Indiana University, Bloomington.

Greiner, R. and Stanley, O. (2013) 'More than money for conservation: Exploring social co-benefits from PES schemes', *Land Use Policy*, vol 31, pp.4–10.

Grove, R. H. (1997) *Ecology, Climate and Empire: Colonialism and Global Environmental History, 1400–1940*, White Horse Press, Cambridge.

Guyer, J. (2004) *Marginal Gains: Monetary Transactions in Atlantic Africa*, University of Chicago Press, Chicago.

Guyer, J. (2009) 'Composites, fictions and risk: Towards an ethnography of price', in K. Hart and C. Hann (eds) *Market and Society: The Great Transformation Today*, Cambridge University Press, Cambridge, pp.203–220.

Hall, A. (2007) 'Social policies in the World Bank: Paradigms and challenges', *Global Social Policy* vol 7, no 2, pp.151–175.

Hall, D., Hirsch, P. and Li, T. (2011) *Powers of Exclusion: Land Dilemmas in Southeast Asia*, University of Hawaii Press, Honolulu.

Hall, R. (2008) *REDD Myths: A Critical Review of Proposed Mechanisms to Reduce Emissions from Deforestation and Degradation in Developing Countries*, issue 114, Friends of the Earth International, www.foei.org/resources/publications/publications-by-subject/forests-and-biodiversity-publications/redd-myths/, accessed 30 August 2014.

Hamilton, K., Bayon, R., Turner, G. and Higgins, D. (2007) *State of the Voluntary Carbon Markets 2007. Picking Up Steam*, The Ecosystem Marketplace and New Carbon Finance, Washington DC and London.

Hamilton, K., Sjardin, M., Marcello, T. and Xu, G. (2008) 'Forging a frontier: State of the voluntary carbon markets 2008', Ecosystem Market Place and New Carbon Finance, San Francisco and London.

Hansen, T. B. and Stepputat, F. (eds) (2001) *States of Imagination: Ethnographic Explorations of the Postcolonial State*, Duke University Press, Durham and London.

Harvard (2012) 'STS research platform: Sociotechnical imaginaries', Harvard Kennedy School, http://sts.hks.harvard.edu/research/platforms/imaginaries/, accessed 26 August 2014.

Harvey, D. (2006) *Spaces of Global Capitalism: A Theory of Uneven Geographical Development*, Verso, London.

Hashmiu, I. (2012) *Carbon Offsets and Agricultural Livelihoods: Lessons Learned from a Carbon Credit Project in the Transition Zone of Ghana*, STEPS Working Paper, no 50, STEPS Centre, Brighton.

Hawthorne, W. D. (1994) 'Fire damage and forest regeneration in Ghana. Forestry inventory and management project of the Ghana Forestry Department', *ODA Forestry Series*, no 4, pp.1–18.

Hemp, C. and Hemp, A. (2008) 'The Chagga Home gardens on Kilimanjaro: Diversity and refuge function for indigenous fauna and flora in anthropogenically influenced habitats in tropical regions under global change on Kilimanjaro, Tanzania', *IHDP Update* 2, https://www.ehs.unu.edu/file/get/7173.pdf, accessed 22 August 2014.

Higuchi, K. (2009) 'Analysis of political economy of conflict and development policies in Sierra Leone: Institutional development for greed and social development for grievance', research paper for the degree of Master of Arts in Development Studies, Graduate School of Development Studies, International Institute of Social Studies, The Hague, The Netherlands.

Hobsbawm, E. and Ranger, T. (eds) (1983) *The Invention of Tradition*, Cambridge University Press, Cambridge.

Holmes, T. and Scoones, I. (2000) *Participatory Environmental Policy Processes: Experiences from North and South*, IDS Working Paper, no 113, Institute of Development Studies, Brighton.

Hughes, D. M. (2006a) 'Whites and water: How Euro-Africans made nature at Kariba Dam', *Journal of Southern African Studies*, vol 32, no 4, pp.823–838.

Hughes, D. M. (2006b) 'Hydrology of hope: Farms, dams, conservation, and whiteness in Zimbabwe', *American Ethnologist*, vol 33, no 2, pp.269–287.

Humphreys, M., Sachs, J. and Stiglitz, J. E. (eds) (2007) *Escaping the Resource Curse*, Columbia University Press, New York.

Hutton, J., Adams, W. and Murombedzi, J. (2005) 'Back to barriers? Changing narratives in biodiversity conservation', *Forum for Development Studies*, vol 32, pp.341–370.

Igoe, J. (2010) 'The spectacle of nature in the global economy of appearances: Anthropological engagements with the spectacular mediations of transnational conservation', *Critique of Anthropology*, vol 30, no 4, pp.375–397.

Igoe, J. and Brockington, D. (2007) 'Neoliberal conservation: A brief introduction', *Conservation and Society*, vol 5, no 4, p.432.

Igoe, J., Neves, K. and Brockington, D. (2010) 'A spectacular eco-tour around the historic bloc theorising the convergence of biodiversity conservation and capitalist expansion', *Antipode*, vol 42, no 3, pp.486–512.

Ingram, J., Stevens, T., Clements, T., Hatchwell, M., Krueger, L., Victurine, R., Holmes, C. and Wilkie, D. (2009) *WCS REDD Project Development Guide*, TRANSLINKS, Wildlife Conservation Society and USAID, Washington DC.

Integrated Land Use Assessment (ILUA) (2008) *Integrated Land Use Assessment Zambia 2005–2008*, Ministry of Tourism Environment and Natural Resources, Lusaka.

International Panel for Climate Change (IPCC) (2007) *Land Use Change Forestry Practices – Fourth Assessment Report*, Working Group 3, Cambridge University Press, Cambridge and New York.

International Union for the Conservation of Nature (IUCN) (2014) 'Forest programmes – our work' www.iucn.org/about/work/programmes/forest/fp_our_work/fp_projects/fp_our_work_ll/, accessed 26 August 2014.

Jagger, P., Luckert, M. M. K., Duchelle, A. E., Lund, J. F. and Sunderlin, W. D. (2014) 'Tenure and forest income: Observations from a global study on forests and poverty', *World Development*, DOI: 10.1016/j.worlddev.2014.03.004.

Jasanoff, S., Kim, S-H., and Sperling, S. (2007) *Sociotechnical Imaginaries and Science and Technology Policy: A Cross-National Comparison*, http://stsprogram.org/admin/files/imaginaries/NSF-imaginaries-proposal.pdf, accessed 26 August 2014.

Jindal, R., Swallow, B. and Kerr, J. (2008) 'Forestry-based carbon sequestration projects in Africa: Potential benefits and challenges', *Natural Resources Forum*, vol 32, no 2, pp.116–130.

Johnston, H. H. (1886) *The Kilimanjaro Expedition*, Kegan, Paul and Trench, London.

Kama, K. (2014) 'On the borders of the market: EU emissions trading, energy security, and the technopolitics of "carbon leakage"', *Geoforum*, vol 51, pp.202–212.

Kanninen, M., Murdiyarso, D., Seymour, F., Angelsen, A., Wunder, S. and German, L. (2007) *Do Trees Grow on Money? The Implications of Deforestation Research for Policies to Promote REDD*, Center for International Forestry Research (CIFOR), Bogor, Indonesia.

Kariba REDD+ Project (2012) Project Design Document, Zurich, Switzerland.

Karim, A. B., Kamara, S., Okoni-Williams, A., Kaiwa, F. J., Lamin-Samu, A. and Sam, M. B. (2013) 'Biodiversity survey of the Western Area Peninsula forest reserve', final report, submitted to Deutsche Welthungerhilfe, Sierra Leone, Department of Biological Sciences, Fourah Bay College, University of Sierra Leone, July.

Karsenty, A. (2008) 'The architecture of proposed REDD schemes after Bali: Facing critical choices', *International Forestry Review*, vol 10, no 3, pp.443–457.

Kasanga, R. K. and Kotey, N. A. (2001) *Land Management in Ghana: Building on Tradition and Modernity*, IIED, London.

Keeley, J. and Scoones, I. (2003) *Understanding Environmental Policy Processes: Cases from Africa*, Earthscan, London.

Kelsall, T. (2013) *Business, Politics, and the State in Africa: Challenging the Orthodoxies on Growth and Transformation*, Zed Books, London.

Kenya Agricultural Carbon Project (KACP) (2008) *Project Design Document for the World Bank Carbon Finance Business*, The World Bank Carbon Finance Unit, Washington DC.

Kenya Agricultural Carbon Project (KACP) (2010) *Environmental and Social Assessment for the Kenya Agricultural Carbon Project*, ECF Consultants, KACP Technical series, Kisumu, Kenya.

Kenya Meteorological Department (2012) 1975–2012 Precipitation data for 37 meteorological stations in Kenya, Kenya Meteorological Department, Headquarters, Nairobi, Kenya.

Kenya National Bureau of Statistics (2007) *Kenya Integrated Household Budget Survey 2005/2006*, Government Printer, Nairobi, Kenya.

Kenya National Bureau of Statistics (2009) *Highlights for the 2009 Population Census*, Government Printer, Nairobi, Kenya.

Kijazi, M. H. (2007) *Stakeholder-Centered Forest Evaluations: Needs, Priorities and Wellbeing of Forest Beneficiaries, Kilimanjaro, Tanzania*, PhD Thesis, University of Toronto.

Kijazi, M. H. (forthcoming 2014) *Triad of Exclusion: Resources, Rents and Rivalry in Mt. Kilimanjaro Conservation Estate*, Working Paper, CODESRIA, Dakar.

Kivumbi, C. O. and Newmark, W. D. (1991) 'The history of the half-mile forest strip on Mount Kilimanjaro', in W. D. Newmark (ed) *The Conservation of Mount Kilimanjaro*, IUCN, Gland, Switzerland and Cambridge, UK, pp.81–86.

Knox-Hayes, J. (2013) 'The spatial and temporal dynamics of value in financialization: Analysis of the infrastructure of carbon markets', *Geoforum*, vol 50, pp.117–128.

Korchinsky, M., Freund, J., Cowan, L. and Dodson, R. (2008) 'The Kasigau Corridor REDD+ Project Phase II – Rukinga Sanctuary', project design document (PDD), Wildlife Works, Rukinga.

Koroma, D. S., Turay, A. B. and Moigua, M. B. (2006) *Republic of Sierra Leone 2004 Population and Housing Census. Analytical Report on the Population Projection for Sierra Leone*, UNFPA, Freetown, Sierra Leone.

Kosoy, N. and Corbera, E. (2010) 'Payments for ecosystem services as commodity fetishism', *Ecological Economics*, vol 69, no 6, pp.1228–1236.

Lambrechts, C.,Woodley, B., Hemp, A., Hemp, C. and Nnyiti, P. (2002) *Aerial Survey of the Threats to Mt. Kilimanjaro Forests*, UNDP, UNOPS, UNEP, Kenya Wildlife Society and University of Bayreuth.

Lan, D. (1985) *Guns and Rain: Guerrillas and Spirit Mediums in Zimbabwe*, California University Press, California.

Lane, R. and Stephan, B. (eds) (2014) *The Politics of Carbon Markets*, Routledge, London.

Lansing, D. M. (2012) 'Performing carbon's materiality: The production of carbon offsets and the framing of exchange', *Environment and Planning – Part A*, vol 44, no 1, pp.204–220.

Larson, A. M., Brockhaus, M., Sunderlin, W. D., Duchelle, A., Babon, A., Dokken, T., Pham, T. T., Resosudarmo, I. A. P., Selaya, G., Awono, A. and Huynh, T. B. (2013) 'Land tenure and REDD+: The good, the bad and the ugly', *Global Environmental Change*, vol 23, no 3, pp.678–689.

Larson, A. M. and Petkova, E. (2011) 'An introduction to forest governance, people and REDD+ in Latin America: Obstacles and opportunities', *Forests*, vol 2, no 4, pp.86–111.

Lasco, R. L., Evangelista, R. S. and Pulhin, F. B. (2010) 'Potential of community-based forest management to mitigate climate change in the Philippines', *Small-scale Forestry*, vol 9, no 4, pp.429–443.

Leach, M. and Mearns, R. (eds) (1996) *The Lie of the Land: Challenging Received Wisdom on the African Environment*, James Currey, Oxford.

Leach, M. and Scoones, I. (2013) 'Carbon forestry in West Africa: The politics of models, measures and verification processes', *Global Environmental Change*, vol 23, no 5, pp.957–967.

Leach, M., Scoones, I. and Stirling, A. (2010) *Dynamic Sustainabilities*, Earthscan, London.

Lederer, M. (2012a) 'Market making via regulation: The role of the state in carbon markets', *Regulation & Governance*, vol 6, no 4, pp.524–544.

Lederer, M. (2012b) 'The practice of carbon markets', *Environmental Politics*, vol 21, no 4, pp.640–656.

Lewis, D. and Mosse, D. (eds) (2006) *Development Brokers and Translators: The Ethnography of Aid and Agencies*, Kumarian Press, Bloomfield, CT.

Li, T. M. (2007) 'Practices of assemblage and community forest management', *Economy and Society*, vol 36, no 2, pp.263–293.

Linacre, N., Kossoy, A. and Ambrosi, P. (2011) *State and Trends of the Carbon Market 2011*, World Bank, Washington DC.

Lohmann, L. (2006) 'Carbon trading', *Development Dialogue*, vol 48, pp.31–218.

Lohmann, L. (2009) 'Neoliberalism and the calculable world: The rise of carbon trading', in K. Birch and V. Mykhnenko (eds) *The Rise and Fall of Neoliberalism*, Zed Books, London.

Lovell, H., and MacKenzie, D. (2011) 'Accounting for carbon: The role of accounting professional organisations in governing climate change', *Antipode*, vol 43, no 3, pp.704–730.

Luig, U. and van Oppen, A. (1997) 'Landscape in Africa: Process and vision', *Paideuma*, vol 43, pp.7–45.

Lund, C. (2008) *Local Politics and the Dynamics of Property in Africa*, Cambridge University Press, Cambridge.

Lund, C. and Boone, C. (2013) 'Introduction: Land politics in Africa – Constituting authority over territory, property and persons', *Africa*, vol 83, no 1, pp.1–13.

Lungu, G. (1985) *Administrative Decentralization in the Zambian Bureaucracy*, Zambian Papers, no 18, Institute for African Studies, University of Zambia, Lusaka.

Luttrell, C., Loft, L., Gebara, M. F., Kweka, D., Brockhaus, M., Angelsen, A. and Sunderlin, W. D. (2013) 'Who should benefit from REDD+? Rationales and realities', *Ecology and Society*, vol 18, no 4, p.52.

Lyons, K. and Westoby, P. (2014) 'Carbon colonialism and the new land grab', *Journal of Rural Studies*, vol 36, no 3, pp.13–21.

MacKenzie, D. (2009) 'Making things the same: Gases, emission rights and the politics of carbon markets', *Accounting, Organizations and Society*, vol 24, no 3–4, pp.440–455.

MacKenzie, D. (2010) 'Constructing carbon markets: Learning from experiments in the technopolitics of emissions trading schemes', in A. Lakoff (ed) *Disaster and the Politics of Intervention*, Columbia University Press, New York, pp.130–148.

MacKenzie, D., Muniesa, F. and Siu, L. (eds) (2007) *Do Economists Make Markets? On The Performativity of Economics*, Princeton University Press, Princeton NJ.

Magnani L. and Nersessian, N. (eds) (2009) *Model-Based Reasoning: Science, Technology, Values*, Kluwer & Plenum, New York.

Mamdani, M. (1987) 'Extreme but not exceptional: Towards an analysis of the agrarian question in Uganda', *Journal of Peasant Studies*, vol 14, no 2, pp.191–225.

Marino, E. and Ribot, J. (2012) 'Special issue introduction: Adding insult to injury: Climate change and the inequities of climate intervention', *Global Environmental Change*, vol 22, no 2, pp.323–328.

Matakala, P., Kokwe, M. and Statz, J. (2014) *Issues and Options Report. Towards a REDD+ Strategy in Zambia*, UN-REDD, Lusaka.

May, P. H., Boyd, E., Veiga, F., Manyu, C., Veiga, F. and Chang, M. (2004) *Local Sustainable Development Effects of Forest Carbon Projects in Brazil and Bolivia: A View from the Field*, IIED, London.

Mbow, C., Smith, P., Skole, D., Duguma, L. and Bustamante, M. (2014) 'Achieving mitigation and adaptation to climate change through sustainable agroforestry practices in Africa', *Current Opinion in Environmental Sustainability*, vol 6, pp.8–14.

McAfee, K. (1999) 'Selling nature to save it? Biodiversity and green developmentalism', *Environment and Planning D: Society and Space*, vol 17, no 2, pp.133–154.

McAfee, K. (2012) 'The contradictory logic of global ecosystem services markets', *Development and Change*, vol 43, no 1, pp.105–131.

McGregor, J. (2009) *Crossing the Zambezi: The Politics of Landscape on a Central Africa Frontier*, James Currey, London.

Mehta, L., Leach, M., Newell, P., Scoones, I., Sivaramakrishnan, K. and Way, S. (1999) *Exploring Understandings of Institutions and Uncertainty: New Directions in Natural Resource Management*, IDS Discussion Paper, no 372, Institute of Development Studies, Brighton.

Mercer, B., Finighan, J., Sembres, T. and Schaefer, J. (2011) 'Protecting and restoring forest carbon in tropical Africa: A guide for donors and funders', Forests Philanthropy Action Network (FPAN), http://files.forestsnetwork.org/Chapter3.pdf, accessed 26 August 2014.

Metcalfe, S. (2003) 'Impacts of transboundary protected areas on local communities in three Southern African initiatives', paper prepared for the workshop on Transboundary Protected Areas in the Governance Stream of the 5th World Parks Congress, Durban, South Africa, 12–13 September.

Metz, B., Davidson, O. R., Bosch, P. R., Dave, R. and Meyer, L. A. (eds) (2007) *Contribution of Working Group III to the Fourth Assessment Report of the Intergovernmental Panel on Climate Change, 2007*, Chapter 8, Cambridge University Press, Cambridge and New York.

Mickels-Kokwe, G. (1998) *Maize, Market and Livelihoods. A Study of the Agrarian History of Luapula Province, 1950–1995*, Interkont Studies, no 9, Institute of Development Studies, University of Helsinki, Helsinki.

Mickels-Kokwe, G., Kokwe, M. and Zulu, D. (forthcoming) *The Community Dynamics of Carbon Projects in Zambia*, IDS Working Paper, Institute of Development Studies, Brighton.

Millenium Ecosystem Assessment (MA) (2005) *Ecosystems and Human Well-Being: A Synthesis*, Island Press, Washington DC.

Minang, P. A., Bernard, F., van Noordwijk, M. and Kahurani, E. (2011) *Agroforestry in REDD+: Opportunities and Challenges*, ASB Policy Brief, no 26, www.asb.cgiar.org/PDFwebdocs/ASB_PB26.pdf, accessed 9 September 2014.

Ministry of Natural Resources and Tourism (MNRT) (2003) *Resource Economic Analysis of Catchment Forest Reserves in Tanzania*, MNRT, Dar es Salaam.

Mitchell, T. (2007) 'The properties of markets', in D. MacKenzie, F. Muniesa and L. Siu (eds) *Do Economists Make Markets? On the Performativity of Economics*, Princeton University, Princeton NJ, pp.244–75.

Mitchell, W. J. T. (1994) *Landscape and Power*, Chicago University Press, Chicago.

Mongabay (2012) 'The Afro-tropical realm', http://rainforests.mongabay.com/20afro tropical.htm, accessed 26 August 2014.

Moninger, J. (2011) *Report on Re-Demarcation of the Western Area Peninsula Forest Reserve*, Welthungerhilfe and ENFORAC, Freetown.

Moninger, J. (2014) *Conservation of the Sierra Leonean Western Area Peninsula Forest reserve (WAPFoR) and its Watershed*, Final Narrative Report, on behalf of Welthungerhilfe, Bonn, ED:DCI-ENV/2008/15386.

Morgan, M. and Morrison, M. (eds) (1999) *Models as Mediators: Perspectives on Natural and Social Science*, Cambridge University Press, Cambridge.

Morgan, M. S. (2009) *The World in the Model*, Cambridge University Press, Cambridge.

Mouffe, C. (2005) *On the Political*, Taylor & Francis, London.

Moyo, D. (2009) *Dead Aid: Why Aid is Not Working and How There is Another Way for Africa*, Allen Lane, London.

Mugyenyi, O., Twesigye, B. and Muhereza, E. (2005) *Balancing Nature Conservation and Livelihoods: A Legal Analysis of the Forestry Evictions by the National Forestry Authority*, ACODE Policy Briefing Series, no 13, Kampala.

Mumba, J. (2013) 'Lessons from the local level support for decentralization in Northern Province', Irish Aid a Local Development Programme, unpublished.

Munro, P. G. (2009) 'Deforestation: Constructing problems and solutions on Sierra Leone's Freetown Peninsula', *Journal of Political Ecology*, vol 16, pp.104–122.

Murdiyarso, D., Brockhaus, M., Sunderlin, W. D. and Verchot, L. (2012) 'Some lessons learned from the first generation of REDD+ activities', *Current Opinion in Environmental Sustainability*, vol 4, no 6, pp.678–685.

Murombedzi, J. C. (1992) 'Decentralisation or recentralisation? Implementing CAMPFIRE in the Omay communards of Nyaminyami district of Zimbabwe', PhD Thesis, Centre for Applied Social Sciences, University of Zimbabwe.

Murphree, M. W. (1993) 'Communities as resource management institutions', *IIED Gatekeepers Series*, no 36, IIED, London.

Murwira, A. (2003) 'Scale matters', PhD Thesis, Natural Resources Department, ITC and Wageningen, The Netherlands.

Mutimba, S. (2013) *Scaling up Fuel Efficient Technologies for Domestic, Institutions and Industrial Use with Carbon Benefits in the Kilimanjaro Region*, Final Report, November, RAS Office, Moshi.

Mutimba, S. and Kibulo, R. (2013) 'Scaling up fuel efficient technologies for domestic, institutions and industrial use with carbon benefits in the Kilimanjaro region', workshop presentation, 8 November, SLM-Kilimanjaro, Moshi.

Nakakaawa, C. A., Vedeld, P. O. and Aune, J. B. (2010) 'Spatial and temporal land use and carbon stock changes in Uganda: Implications for a future REDD strategy', *Mitigation and Adaptation Strategies for Global Change*, vol 16, no 1, pp.25–62.

Nashanda, E. (2013) *UN-REDD Tanzania Programe*, presented to the UN-REDD 11th Policy Board Meeting, 9 November.

National Forestry Authority (2005) Uganda's Forests, Functions and Classification, http://nfa.org.ug/docs/forests_functions_and_classification.pdf, accessed 20 August 2014.

National Land Alliance (2007) 'National Land Policy: The popular version', Kenya Land Alliance, Nakuru.

Naughton-Treves, L., Wendland, K., Alix-Garcia, J., Baird I. and Turner, M. (eds) (2014) 'Land tenure and tropical forest carbon management', *World Development*, vol 55, pp.1–84.

Nel, A. (forthcoming) 'Market environmentalism and reterritorialization: A critical geography of carbon forestry in Uganda', *Journal of Political Ecology*.

Nel, A. and Hill, D. (2013) 'Constructing walls of carbon – the complexities of community, carbon sequestration and protected areas in Uganda', *Journal of Contemporary African Studies*, vol 31, no 3, pp.421–440.

Nelson, F. (ed) (2010) *Community rights, Conservation and contested land: the politics of natural resource governance in Africa*, London, Earthscan.

Neumann, R. P. (1998) *Imposing Wilderness: Struggles over Livelihood and Nature Preservation in Africa*, University of California Press, Berkley.

New Forests Company (2011) *New Forests Company: Response to Oxfam*, www.newforests.net/index.php/responsibility/response-to-oxfam, accessed 13 December 2011. The initial link has been removed and the remaining statement can be found here: www.newforests.net/index.php/hmd_article/statement-from-the-new-forests-company-regarding-the-oxfam-report, accessed 7 October 2014.

New Forests Company (2014) New Forests Company, Uganda, www.newforests.net/index.php/hmd_article/team-management, accessed 4 October 2014.

Newell, P., Boykoff, M. and Boyd, E. (eds) (2012) *The New Carbon Economy: Constitution, Governance and Contestation*, vol 22, John Wiley, Chichester.

Nixon, R. (2011) *Slow Violence*, Harvard University Press, Cambridge MA.

Nsiah-Gyabaah, K. (1996) 'Bush fires in Ghana', IFFN Country Report, vol 15, pp.1–10.

Nyaoro, W. (2001) 'Choice and use of rural water supply systems: Environmental and socio-cultural dimensions', in M. A. Mohamed Salih (ed) *Local Environmental Change and Society in Africa*, Kluwer Academic Publishers, Dordrecht, The Netherlands.

Odner, K. (1971) 'A preliminary report on an archaeological survey on the slopes of Kilimanjaro', *Azania*, vol 6, pp.131–150.

Offen, K. H. (2004) 'Historical political ecology: An introduction', *Historical Geography*, vol 32, pp.19–42.

Olander, J. and Ebeling, J. (2011) *Building Forest Carbon Projects: Step-by-Step Overview and Guide*, Forest Trends/EcoDecisión, Washington DC.

Österreichische Bundesforste (OBF) (2012) *REDD+ Scoping Study for the Western Area Peninsula Forest Reserve,* Final Report submitted to 'Conservation of the Sierra Leone Western Area Peninsula Forest Reserve and its Watersheds' Project, Purkersdorf, Austria, OBF.

Oxfam America (2008) *Turning Carbon into Gold – Oxfam America*, Oxfam America, Washington DC.

Pagiola, S., Arcenas, A. and Platais, G. (2005) 'Can payments for environmental services help reduce poverty? An exploration of the issues and the evidence to date from Latin America', *World Development*, vol 33, no 2, pp.237–253.

Palm, C. A., Vosti, S. A., Sanchez, P. A. and Ericksen, P. J. (eds) (2013) *Slash-and-Burn Agriculture: The Search for Alternatives*, Columbia University Press, New York.

Parrotta, J. A., Wildburger, C. and Mansourian, S. (eds) (2012) *Understanding Relationships between Biodiversity, Carbon, Forests and People: The Key to Achieving REDD+ Objectives*, International Union of Forest Research Organizations (IUFRO), Vienna.

Pearson, T., Walker, S., Chalmers, J., Swails, E. and Brown, S. (2009) *Guidebook for the Formulation of Afforestation/Reforestation and Bioenergy Projects in the Regulatory Carbon Market*, Winrock International, Arlington VA.

Peet, R., Robbins, P. and Watts, M. (eds) (2010) *Global Political Ecology*, Taylor & Francis, London.

Peluso, N. L. and Lund, C. (2011) 'New frontiers of land control: Introduction', *Journal of Peasant Studies*, vol 38, no 4, pp.667–681.

Peskett, L., Schreckenberg, K. and Brown, J. (2011) 'Institutional approaches for carbon financing in the forest sector: Learning lessons for REDD+ from forest carbon projects in Uganda', *Environmental Science & Policy*, vol 14, no 2, pp.216–229.

Peters, P. E. (2004) 'Inequality and social conflict over land in Africa', *Journal of Agrarian Change*, vol 4, no 3, pp.269–314.

Peters, P. E. (2009) 'Challenges in land tenure and land reform in Africa: Anthropological contributions', *World Development*, vol 37, no 8, pp.1317–1325.

Peters-Stanley, M. and Gonzalez, G. (2014) *Sharing the Stage: State of the Voluntary Carbon Markets 2014 (Executive Summary)*, Forest Trends' Ecosystem Marketplace, Washington DC, www.forest-trends.org/documents/files/doc_4501.pdf, accessed 30 August 2014.

Peters-Stanley, M., Gonzalez, G. and Yin, D. (2013) *Covering New Ground: State of the Forest Carbon Markets 2013 Report*, Forest Trends' Ecosystem Marketplace, Washington DC.

Peters-Stanley, M., Hamilton, K., Marcelo, T. and Sjardin, M. (2011) *Back to the Future: State of the Voluntary Carbon Markets 2011*, Ecosystem Marketplace and Bloomberg New Energy Finance, Washington, DC, www.forest-trends.org/documents/files/doc_2828.pdf, accessed 7 October 2014.

Peters-Stanley, M. and Yin, D. (2013) *Maneuvering the Mosaic. State of the Voluntary Carbon Markets 2013*, Forest Trends' Ecosystem Marketplace and Bloomberg New Energy Finance, Washington DC.

Pham, T. T., Brockhaus, M., Wong, G., Le, N. D., Tjajadi, J. S., Loft, L., Luttrell, C. and Assembe Mvondo, S. (2013) *Approaches to Benefit Sharing: A Preliminary Comparative Analysis of 13 REDD+ Countries*, Center for International Forestry Research (CIFOR), Bogor, Indonesia.

Pokorny, B., Scholz, I. and de Jong, W. (2013) 'REDD+ for the poor or the poor for REDD+? About the limitations of environmental policies in the Amazon and the potential of achieving environmental goals through pro-poor policies', *Ecology and Society*, vol 18, no 2, p.3.

Portaccio, A., Pettenella, D. and Brotto, L. (2013) 'Endorsing REDD+ in the institutional mechanisms: How could tropical forests and voluntary initiatives be affected?', http://intra.tesaf.unipd.it/pettenella/index.html, accessed 30 August 2014.

REDD-monitor (2011) 'Can financial markets solve the climate crisis?' www.redd-monitor.org/2011/04/01/can-financial-markets-solve-the-climate-crisis/, accessed 3 August 2011.

Redford, K. and Adams, W. (2009) 'Payment for ecosystem services and the challenge of saving nature', *Conservation Biology*, vol 23, pp.785–787.

Regional Administrative Secretary (RAS) – Kilimanjaro (2013) *Kilimanjaro Sustainable Land Management*, Moshi.

Republic of Kenya (2008) *Siaya District Development Plan for 2008–2012*, Government Printer, Nairobi, Kenya.

Republic of Kenya (2010a) *Agricultural Sector Development Strategy for 2010–2020*, Government Printer, Nairobi, Kenya.

Republic of Kenya (2010b) *Kenya National Climate Change Response Strategy*, Government Printer, Nairobi, Kenya.

Republic of Sierra Leone (RoSL), Ministry of Agriculture, Forestry and Food Security (MAFFS) and EU (2014) 'REDD+ and capacity building project in Sierra Leone', Direct Decentralized Operation, Global Commitment: DCI ENV 2011/023261 Operational Programme Estimate No 1 for Operational Period from March 2014 to March 2015.

Ribot, J. C. and Larson, A. M. (2012) 'Reducing REDD risks: Affirmative policy on an uneven playing field', *International Journal of the Commons*, vol 6, no 2, pp.233–254.

Ribot, J. C. and Peluso, N. L. (2003) 'A theory of access', *Rural Sociology*, vol 68, no 2, pp.153–181.

Ribot, J. C., Agrawal, A. and Larson, A. M. (2006) 'Recentralizing while decentralizing: How national governments reappropriate forest resources', *World Development*, vol 34, no 11, pp.1864–1886.

Richards, P. (1985) *Indigenous Agricultural Revolution: Ecology and Food Production in West Africa*, Hutchinson, London.

Richards, P. (1996) *Fighting for the Rain Forest: War, Youth and Resources in Sierra Leone*, James Currey, Oxford.

Rivera, L. (2014) 'Introduction to the UNFCCC REDD+ safeguards', REDD+ safeguards capacity-building workshop, Ndola, Zambia, 17 September.

Robbins, P. (2012) *Political Ecology: A Critical Introduction*, Wiley-Blackwell, Malden MA.

Rockström, J., Steffen, W., Noone, K., Persson, Å., Chapin, III, F. S., Lambin, E. F., Lenton, T. M., Scheffer, M., Folke, C., Schellnhuber, H. J., Nykvist, B., de Wit, C. A., Hughes, T., van der Leeuw, S., Rodhe, H., Sörlin, S., Snyder, P. K., Costanza, R., Svedin, U., Falkenmark, M., Karlberg, L., Corell, R. W., Fabry, V. J., Hansen, J., Walker, B., Liverman, D., Richardson, K., Crutzen, P. and Foley, J. A. (2009) 'A safe operating space for humanity', *Nature*, vol 461, pp.472–475.

Roe, E. M. (1991) 'Development narratives, or making the best of blueprint development' *World Development*, vol 19, no 4, pp.287–300.

Rutherford, B, (2002) *Working on the Margins: Black Workers, White Farmers in Postcolonial Zimbabwe*, Weaver Press, Harare.

Sachs, J. D. and Warner, A. M. (1995) *Natural Resource Abundance and Economic Growth*, NBER Working Paper 5398, National Bureau of Economic Research, Cambridge MA.

SAH/D (2007) *Land, Agricultural Change and Conflict in West Africa Regional Issues from Sierra Leone, Liberia and Côte d'Ivoire: Historical Overview*, Transformation du monde rural et Developpement durable en Afrique de L'Ouest, Transformation and Sustainable Development in West Africa, Club du Sahelet de L'Afriques de L'Ouest (CSAO), Sahel and West Africa Club (SWAC) OCOE.

Sassen, S. (2013) 'Land grabs today: Feeding the disassembling of national territory', *Globalizations*, vol 10, no 1, pp.25–46.

Sayer, J. A., Harcourt, C. S. and Collins, N. M. (1992) *Conservation Atlas of Tropical Forests: Africa*, Macmillan, London.

Sayer J. A., Sunderland, T. C. H., Ghazoul, J., Pfund, J. L., Sheil, D., Meijard, E., Venter, M., Boedhihartono, A. K., Day, M., García, C., Van Oosten, C. and Buck, L. E. (2013)

'Ten principles for a landscape approach to reconciling agriculture, conservation, and other competing land uses', *Proceedings of the National Academy of Sciences* (PNAS), vol 110, no 21.

Schroeder, H. and McDermott, C. (2014) 'Beyond carbon: Enabling justice and equity in REDD+ across levels of governance', *Ecology and Society*, vol 19, no 1, p.31.

Scoones, I. (ed) (1995) *Living with Uncertainty: New Directions in Pastoral Development in Africa*, IT Publications, London.

Scoones, I. (2009) 'Livelihoods perspectives and rural development', *Journal of Peasant Studies*, vol 36, no 1, pp.171–196.

Scoones, I., Marongwe, N., Mavedzenge, B., Murimbarimba, F., Mahenehene, J. and Sukume, C. (2011) 'Zimbabwe's land reform: Challenging the myths', *Journal of Peasant Studies*, vol 38, pp.967–993.

Scoones, I. and Thompson, J. (1994) *Beyond Farmer First: Rural People's Knowledge, Agricultural Research and Extension Practice*, IT Publications, London.

Scott, J. C. (1985) *Weapons of the Weak: Everyday Forms of Peasant Resistance*, Yale University Press, New Haven.

Scott, J. C. (1990) *Domination and the Arts of Resistance: Hidden Transcripts*, Yale University Press, New Haven.

Scott, J. C. (2009) *The Art of Not Being Governed: An Anarchist History of Upland Southeast Asia*, Yale University Press, New Haven and London.

Seifert-Granzin, J. (2011) 'REDD guidance: Technical project design', in J. Ebeling and J. Olander (eds) *Building Forest Carbon Projects*, Forest Trends, Washington DC.

Sibley, D. (1995) *Geographies of Exclusion: Society and Difference in the West*, Routledge, London.

Sikor, T., Auld, G., Bebbington, A. J., Benjaminsen, T. A., Gentry, B. S., Hunsberger, C., Izac, A.-M., Margulis, M. E., Plieninger, T., Schroeder, H. and Upton, C. (2013) 'Global land governance: From territory to flow?', *Current Opinion in Environmental Sustainability*, vol 5, no 5, pp.522–527.

Sikor, T. and Lund, C. (eds) (2010) *The Politics of Possession: Property, Authority, and Access to Natural Resources*, John Wiley, Chichester.

Silber, T. and von Laer, Y. (2012) *Kariba Carbon+ Project CCBS Project Design Document (PDD)*, South Pole Carbon Asset Management, www.southpolecarbon.com/_downloads/990_CCBA-PDD.pdf, accessed 6 September 2014.

Simula, M. (2010) 'Analysis of REDD+ financing gaps and overlaps', http://reddpluspartnership.org/25159-09eb378a8444ec149e8ab32e2f5671b11.pdf, accessed 30 August 2014.

Sirohi, S. (2007) 'CDM: Is it a 'win–win'strategy for rural poverty alleviation in India?', *Climatic Change*, vol 84, no 1, pp.91–110.

Spash, C. L. (2010) 'The brave new world of carbon trading', *New Political Economy*, vol 15, no 2, pp.169–195.

Spiereneburg, M. (2004) *Strangers, Spirits and Land Reforms: Conflicts about Land in Dande, Northern Zimbabwe*, Brill, Leiden.

Springate-Baginski, O. and Wollenberg, E. (eds) (2010) *REDD, Forest Governance and Rural Livelihoods: The Emerging Agenda*, Center for International Forestry Research (CIFOR), Bogor, Indonesia.

Stephan, B. and Paterson, M. (2012) 'The politics of carbon markets: An introduction', *Environmental Politics*, vol 21, no 4, pp.545–562.

Streck, C. (2012) 'Financing REDD+: Matching needs and ends', *Current Opinion in Environmental Sustainability*, vol 4, pp.628–637.

Stripple, J. and Bulkeley, H. (eds) (2013) *Governing the Climate*, Cambridge University Press, Cambridge.

Sullivan, S. (2011) *Banking Nature? The Financialisation of Environmental Conservation*, Open Anthropology Cooperative Press, Working Papers Series, no 8, http://openanthcoop. net/press/2011/03/11/banking-nature/, accessed 18 October 2014.

Sullivan, S. (2013) 'Banking nature? The spectacular financialisation of environmental conservation', *Antipode*, vol 45, no 1, pp.198–217.

Sunderland, T. C. H., Achdiawan, R., Angelsen, A., Babigumira, R., Ickowitz, A., Paumgarten, F., Reyes-García, V. and Shively, G. (2014) 'Challenging perceptions about men, women, and forest product use: A global comparative study', *World Development*, DOI: 10.1016/j.worlddev.2014.03.003.

Sunderlin, W. D., Ekaputri, A. D., Sills, E. O., Duchelle, A. E., Kweka, D., Diprose, R., Doggart, N., Ball, S., Lima, R., Enright, A., Torres, J., Hartanto, H. and Toniolo, A. (2014a) 'The challenge of establishing REDD+ on the ground: Insights from 23 subnational initiatives in six countries', Center for International Forestry Research (CIFOR), Bogor, Indonesia.

Sunderlin, W. D., Larson, A. M., Duchelle, A. E., Pradnja Resosudarmo, I. A., Huynh, T. B., Awono, A. and Dokken, T. (2014b) 'How are REDD+ proponents addressing tenure problems? Evidence from Brazil, Cameroon, Tanzania, Indonesia, and Vietnam', *World Development*, vol 55, pp.37–52.

Tanzania Daily News (2013) 'Mount Kilimanjaro needs serious protection', Deogratius Mushi, 14 July, http://archive.dailynews.co.tz/index.php/features/19854-mount-kilimanjaro-needs-serious-protection, accessed 15 September 2014.

Tanzania National Parks (TANAPA) (2001) 'Proposal for annexing Mount Meru and Mount Kilimanjaro forest reserves to Arusha and Kilimanjaro national parks', Main Report, TANAPA.

Tanzania National Parks (TANAPA) (2014) 'Number of Climbers on Mt. Kilimanjaro not a threat', Press Statement, 5 March.

Tanzania National Parks (TANAPA) and Institute of Resource Assessment University of Dar es Salaam (IRA-UD) (2002) *Assessment of the Status of Proposed Extension and Establishment of New National Parks and the Surrounding Areas*, TANAPA.

Taylor, C. (2004) *Modern Social Imaginaries*, Duke University Press, Durham and London.

Thomas, S., Dargusch, P., Harrison, S. and Herbohn, J. (2010) 'Why are there so few afforestation and reforestation Clean Development Mechanism projects?', *Land Use Policy*, vol 27, no 3, pp.880–887.

To, P. X., Dressler, W. H., Mahanty, S., Pham, T. T. and Zingerli, C. (2012) 'The prospects for payment for ecosystem services (PES) in Vietnam: A look at three payment schemes', *Human Ecology*, vol 40, no 2, pp.237–249.

TREES (2010) 'Feasibility study AR for the 'Aframso' project in Ghana prepared for African carbon traders, 22 November 2010', TREES Forest Carbon Consulting LLC, Wollerau, Switzerland.

Tsing, A. L. (2005) *Friction: An Ethnography of Global Connection*, Princeton University Press, Oxford.

Turyahabwe, N. and Banana, A.Y. (2008) 'An overview of history and development of forest policy and legislation in Uganda', *International Forestry Review*, vol 10, no 4, pp.641–656.

Ucko, P. J. and Layton, R. (1999) *The Archaeology and Anthropology of Landscape: Shaping Your Landscape*, Routledge, London and New York.

United Nations Collaborative Programme on Reduced Emissions from Deforestation and Degradation (UN-REDD) (2014) UN-REDD Programme Regions and Partner

Countries, www.un-redd.org/Partner_Countries/tabid/102663/Default.aspx, accessed 30 March 2014.

United Nations Environment Programme (UNEP) (2010) 'Pathways for implementing REDD+: Experiences from carbon markets and communities', *Perspective Series*, UNEP Risø Centre, Roskilde, Denmark.

United Nations Environment Programme (UNEP) (2011) *Towards a Green Economy: Pathways to Sustainable Development and Poverty Eradication*, United Nations Environment Programme, Nairobi, Kenya, www.unep.org/greeneconomy, accessed 22 August 2014.

United Nations Environment Programme (UNEP) (2014) 'Landscapes' www.landscapes. org/about/, accessed May 2014.

United Nations Framework Convention on Climate Change (UNFCCC) (2009) 'The Copenhagen Accord', *Report of the Conference of the Parties on its Fifteenth Session*, 7–18 December 2009, Copenhagen (ref: FCCC/CP/2009/L.7).

United Nations Framework Convention on Climate Change (UNFCCC) (2011) *Report of the Conference of the Parties on its Sixteenth Session, Held in Cancun*, 29 November–10 December 2010, (FCCC/CP/2010/7/Add.1), http://unfccc.int/resource/docs/2010/cop16/eng/07a01.pdf, accessed 30 August 2014.

United Nations Framework Convention on Climate Change (UNFCCC) (2012) *Report of the Conference of the Parties on its Seventeenth Session*, 28 November–11 December 2011, Durban (ref: FCCC/CP/2011/9/Add.1).

United Nations Framework Convention on Climate Change (UNFCCC) (2013) *Benefits of the Clean Development Mechanism 2012*, https://cdm.unfccc.int/about/dev_ben/ABC_2012.pdf, accessed 30 August 2014.

United Nations Framework Convention on Climate Change (UNFCCC) (2014) *Report of the Conference of the Parties on its Nineteenth Session*, 11–23 November 2013, Warsaw, Bonn, Germany.

United Republic of Tanzania (URT) (1998) *National Forest Policy*, Government Printer.

United Republic of Tanzania (URT) (2013) *National Strategy for Reduced Emissions from Deforestation and Forest Degradation*, Vice President Office, Dar es Salaam.

Unruh, J. D. (2008) 'Carbon sequestration in Africa: The land tenure problem', *Global Environmental Change*, vol 18, no 4, pp.700–707.

Vaccaro, I., Beltran, O. and Paquet, P. A (2013) 'Political ecology and conservation policies: Some theoretical genealogies', *Journal of Political Ecology*, vol 20, pp.255–272.

Van Bodegom, A. J., Savenije, H. and Wit, M. (eds) (2009) *Forests and Climate Change: Adaptation and Mitigation*, Tropenbos International, Wageningen, www.etfrn.org/publications/forests+and+climate+change:+adaptation+and+mitigation, accessed 28 September 2014.

van Noordwijk, M., Agus, F., Dewi, S., Ekadinata, A., Tata, H. L., Suyanto, S., Galudra, G. and Pradhan, U. P. (2010) *Opportunities for Reducing Emissions from All Land Uses in Indonesia: Policy Analysis and Case Studies, Partnership for the Tropical Forest Margins, Kenya*, www.asb.cgiar.org/PDFwebdocs/REALU%20Indonesia_Final.pdf, accessed 9 September 2014.

Vatn, A. and Angelsen, A. (2009) 'Options for a national REDD+ architecture', in A. Angelsen with M. Brockhaus, M. Kanninen, E. Sills, W. D. Sunderlin and S. Wertz-Kanounnikoff (eds) *Realising REDD+ National Strategy and Policy Options*, Center for International Forestry Research (CIFOR), Bogor, Indonesia, pp.57–74.

Verified Carbon Standard (VCS) (2013a) *Approved VCS Methodology VM0015: Methodology for Avoided Unplanned Deforestation*, http://v-c-s.org/sites/v-c-s.org/files/VM0015%20Methodology%20for%20Avoided%20Unplanned%20Deforestation%20v1.1.pdf, accessed 15 January 2013.

Verified Carbon Standard (VCS) (2013b) http://v-c-s.org/methodologies/find, accessed 15 January 2013.

Verified Carbon Standard (VCS) (2013c) http://v-c-s.org/node/286, accessed 15 January 2013.

Vhugen, D., Aguilar, S. and Miner, S. (2012) *REDD+ and Carbon Rights: Lessons from the Field*, Property Rights and Resource Governance Project (PRRGP), USAID Working Paper, www.landandpoverty.com/agenda/pdfs/paper/vhugen_full_paper.pdf, accessed 20 April 2012.

Virgilio, N. and Marshall, S. (2009) *Forest Carbon Strategies in Climate Change Mitigation: Confronting Challenges through on-the-Ground Experience*, The Nature Conservancy, Arlington.

Vision 2050 Forestry (undated) 'Documentation for tree farming and coding', Vision 2050 Forestry Project Document.

Visseren-Hamakers, I. J., McDermott, C., Vijge, M. J. and Cashore, B. (2012) 'Trade-offs, co-benefits and safeguards: Current debates on the breadth of REDD+', *Current Opinion in Environmental Sustainability*, vol 4, no 6, pp.646–653.

Waern, K. (1984) *The Ward as a Unit for Planning, Administration and Co-Operation on Local Level*, National Commission for Development Planning, Lusaka.

Wekesa, A. W. (2010) 'Kenya Agriculture Carbon Finance Project', PowerPoint presentation.

Welthungerhilfe (WHH) (2014) 'Conservation of the Western Area Peninsula Forest and its watershed in Sierra Leone: Avoided deforestation', project brochure, Freetown www.ecosalone.com/REDD%20opportunities%20in%20the%20WAPFR-WHH.pdf, accessed 22 August 2014.

White, B., Borras, M., Hall, R. and Scoones, I. (2012) 'The new enclosures: A critical perspective on corporate land deals', *Journal of Peasant Studies*, vol 39, no 9, pp.619–647.

White, G. (1993) 'Towards a political analysis of markets', *IDS Bulletin*, vol 24, no 3, pp.4–11.

Wildlife Works (2008) 'The Kasigau Corridor REDD+ Project Phase I – Rukinga Sanctuary', Project Design Document (PDD) for Validation Using the Climate, Community and Biodiversity (CCB), Wildlife Works, Rukinga.

Wildlife Works (2011) 'The Kasigau Corridor REDD+ Project, Phase I, Rukinga Sanctuary' Project Document (PD) for Validation Using the Voluntary Carbon Standard (VCS) 2007.1/Sectoral Scope 14 VM0009, methodology for avoided mosaic deforestation of tropical forests, Version 9, Wildlife Works, Rukinga.

Williams, R. (1973) *The Country and the City*, Oxford University Press, New York.

Winnebah, T. R. A. (2003) *Food Security Situation in Sierra Leone Since 1961*, Food Security Monograph no 2, World Food Programme, Technical Support Unit, Freetown, Sierra Leone.

Winnebah, T. R. A. (2006) *The Tabule Theatre Group Play and The Gola Forest, Sierra Leone's Community's Perception and Protection of Chimpanzees*, Wild Chimpanzee Foundation, Environmental Education Monograph, no 3, June, Freetown, Sierra Leone.

Winnebah, T. R. A. (2008) *Koinadugu District Loma Mountain Area Offset (LMAO)*, socio-economic study, completion of the Bumbuna Falls Hydroelectric Project, Government of Sierra Leone, Ministry of Energy and Power, Freetown, Sierra Leone.

Winnebah, T. R. A. (2010) *West African Forest Strategy: Sierra Leone*, LTS International and ONF International, Contract: 'A West Africa Forest Strategy', Project ID: 100024717.

Winnebah, T. R. A. (2011) *Perception of Chimpanzees and their Protection by the Neighbouring Communities of the Outamba-Kilimi National Park, Sierra Leone and Guinea after the 'Tabule'*

Theatre Tour, Wild Chimpanzee Foundation (WCF), Chimpanzee Environmental Education Monograph, no 6, August, Freetown, Sierra Leone.

Winnebah, T. R. A. and Cofie, O. (2007) 'Urban farms after a war', in *State of the World, Our Urban Planet*, Worldwatch Institute, pp.64–65, www.worldwatch.org/files/pdf/State%20of%20the%20World%202007.pdf, accessed 30 September 2014.

Wood, A. P. (1986) 'Agriculture in Lusaka Province', in G. J. Williams (ed), *Lusaka and its Environs: A Geographical Study of a Planned Capital City in Tropical Africa*, Zambia Geographical Association Handbook Series, no 9, Lusaka, pp.316–331.

Worby, E. (1995) 'Cotton, commoditisation and social form in Gokwe, Zimbabwe', *Journal of Peasant Studies*, vol 23, no 1, pp.1–29.

World Bank (1993) *Ghana 2000 and Beyond: Setting the Stage for Accelerated Growth and Poverty Reduction*, World Bank, Washington DC.

World Bank (2005) *The BioCarbon Fund Overview,* BioCF Project Training Seminar, World Bank, Washington DC.

World Bank (2006) *Ghana Country Environmental Analysis*, Environmentally and Socially Sustainable Development Department (AFTSD), Africa Region, World Bank, Washington DC.

World Bank (2008) *World Development Report 2008: Agriculture for Development*, World Bank, Washington DC.

World Bank (2010a) *The Hague Conference on Agriculture, Food Security and Climate Change: Opportunities and Challenges for a Converging Agenda: Country Examples*, World Bank, Washington DC.

World Bank (2010b) *Project Information Document, Project number 53088*, World Bank, Washington DC.

World Bank (2011) *BioCarbon Fund Experience; Insights from Afforestation and Reforestation Clean Development Mechanism Projects*, Carbon Finance Unit, World Bank, Washington DC.

World Bank (2014) *Landscape Approaches in Sustainable Development*, World Bank, http://web.worldbank.org/WBSITE/EXTERNAL/TOPICS/EXTARD/0,,contentMDK:23219902~pagePK:148956~piPK:216618~theSitePK:336682,00.html, accessed 9 January 2014.

Wunder, S. (2005) *Payments for Environmental Services: Some Nuts and Bolts*, Center for International Forestry Research (CIFOR), Bogor, Indonesia.

WWF (2013) 'Uganda unveils the world's first Earth Hour forest', http://wwf.panda.org/?207595/Uganda-unveils-the-worlds-first-Earth-Hour-forest, accessed 24 September 2014.

Zimbabwe National Statistics Agency (2012) *Census 2012: Preliminary Census Report*, ZIMSTATS, Harare.

Zomer, R. J., Trabucco, A., Coe, R. and Place, F. (2009) *Trees on Farm: Analysis of Global Extent and Geographical Patterns of Agroforestry*, Working Paper no 89, World Agroforestry Centre, Nairobi, Kenya.

INDEX

Locators in *italic* refer to figures and tables

media discourses 1, 28, 37–9

Meeting of the Parties to the Kyoto
  Protocol (CMP) 47

migrants 18; Ghana 37, 164, 173, 175,
  178, 179; Sierra Leone 183, 189–90;
  Uganda 101; Zambia 128;
  Zimbabwe 30, 35, 142, 143, 146–8,
  158–60, 162. *see also* squatters

militaristic approaches. *see* guards/
  policing

Millennium Development Goals
  (MDG), United Nations 61

mineral exploitation 2, 3, 22, 96, 127,
  182, 193

Ministry of Agriculture, Forestry and
  Food Security (MAFFS), Sierra
  Leone 180–1, 183–4

Ministry of Lands, Country Planning
  and the Environment (MLCPE),
  Sierra Leone 183–4

Ministry of Water and Environment
  (MWE), Uganda 96

missionary vision, Zimbabwe 150–2,
  155–6

mitigation strategies 1, 3, 44; Kenya
  91, 108; Sierra Leone 186; Tanzania
  70; Zambia 138–9. *see also* carbon
  offset interventions

MKUKUTA (National Strategy for
  Growth and Reduction of Poverty)
  60

monitoring, reporting and verification
  (MRV) procedures 23, 25, 45;
  Tanzania 60–1; Uganda 98

Mount. Kilimanjaro carbon finance
  project 67–72, 76. *see also* Tanzania
  case study

mushroom growing 37, 71, 165

Mwachusa people, Zambia
  128

narratives 1, 15–17; Ghana 168–72;
  Kenya 112–14; Tanzania 64–7;
  Uganda 99

National Carbon Offset Standard,
  Zimbabwe 148

National Forest Resources Monitoring
  and Assessment approach
  (NAFORMA), Tanzania 61

National Forestry and Tree Planting
  Act, Uganda 95

National Forestry Authority (NFA),
  Uganda 96

national parks *9–11*, 19; Kenya 109,
  111, 112, 114, 115, 121, 122;
  Sierra Leone 184, 186; Tanzania 59,
  61, 63–6, 73; Uganda 96; Zambia
  128, 129; Zimbabwe 145, 146.
  see also protected areas

National Strategy for Growth and
  Reduction of Poverty
  (MKUKUTA) 60

natural capital 2

natural landscapes. *see* pristine
  landscapes

neoliberal economic policies 2, 29, 95,
  96

New Forests Company (NFC), UK
  94, 98, 99, 102–4, 106. *see also*
  Uganda case study

Nile Ply 100

non-governmental organizations
  (NGOs) 3, 25, 39, 47; Kenya 80,
  112; Sierra Leone 180; Zambia 138;
  Zimbabwe 149, 150, 152

non-timber forest products (NTFPs)
  175; Ghana 165, 173, 175, 176, 178;
  Sierra Leone 189; Tanzania 62, 64.
  see also livelihoods

Norway, Green Resources 98, 100, 103

offsets, carbon. *see* carbon offset
  interventions

original forest cover. *see* pristine
  landscapes

Österreichische Bundesforste (OBF)
  187

Oxfam 102, 105, 106

For Product Safety Concerns and Information please contact our EU
representative GPSR@taylorandfrancis.com
Taylor & Francis Verlag GmbH, Kaufingerstraße 24, 80331 München, Germany

www.ingramcontent.com/pod-product-compliance
Lightning Source LLC
Chambersburg PA
CBHW050421280326
41932CB00013BA/1942